Happy 27th Birthday
to my Son

James alvin Winchester

Love from
Dad

July 1976

Fire in America!

Dedicated

to the victims of fire

and to all persons

interested in the science of

fire protection, fire prevention and fire control.

Fire in America!

By Paul Robert Lyons

NFPA Fire Service Editor

Published by
National Fire Protection Association
470 Atlantic Avenue, Boston, Massachusetts 02210

Printed in U.S.A.

First Edition 1976

Standard Book No. 87765–059–4

NFPA No. SPP-33

Library of Congress Card Number 75–29598

Copyright © 1976

National Fire Protection Association

Foreword

This handsome book focuses on a grim aspect of our history on the North American continent; usually dramatic, often heroic, but always, in the final analysis, grim. For all its 272 pages, its stories, its pictures, it tells only a small fraction of the lamentable toll of lives and property that has been taken by fire since the first European colonies were established on these shores.

It is published as a measure of observance of our nation's unique and remarkable history and our startling experience with fire in each century. It is appropriate to this purpose in that it helps us all to realize how far we have come in the last three hundred years, and yet how far we still must go if we are to make life more secure and happiness more successfully pursued.

Fire tends to evoke an emotional response in most people, but it should not stimulate irrational fear and panic. We can control fire. We can live safely with fire. But we must appreciate the full danger of fire and the scope of our national experience with fire. It is the purpose of the publisher of this volume to channel response of readers toward greater concern for fire safety and toward an increased awareness of the need to teach younger generations how to understand the dangers of fire.

Living safely with fire has long been the objective of the many programs, technical and otherwise, which the National Fire Protection Association has been conducting since the turn of the twentieth century. Many of its recent and current programs have been developed specifically for you who now read this book. We hope that this fascinating chronicle of fire in America, will motivate you to seek information on how to make your own life and the lives of those you love more secure from fire at home, at work and at play.

CHARLES S. MORGAN
PRESIDENT
NATIONAL FIRE PROTECTION ASSOCIATION

Table of Contents

Into the Jaws of Death, by C. S. Reinhard. Harper's Weekly, Vol. XVI. Reproduced by permission from the Library of the Boston Athenaeum.

List of Illustrations

Special appreciation is acknowledged to Robert W. Grant, Vice President of the NFPA, who made available century-old newspapers and two valuable books from his private collection. Excerpts from the newspapers will be found in Chapter Two on pages 45, 46, 48, 49, 50, 51, and 55.

Illustrations from *The Story of the New York Volunteer Fire Department* will be found on the following pages: 21, 23, 30, 31, 32, 33, 34, 49, 54, 58, 59, 60, 65, 68, and 70.

Illustrations from *Great Fires in Chicago and the West* will be found on the following pages: 78, 82, 83, 84, 85, 86, 87, and 88.

Special appreciation is also acknowledged to Harold Walker of Marblehead, Massachusetts who made available many pictures of old-time fire apparatus from his personal collection of more than 3,000 photos. Some of these will be found on the following pages: 76, 77, 110, and 111.

Acknowledgements

A book of this type can only be completed with the assistance of many individuals of special talents and knowledge. With deep appreciation, the author acknowledges the contributions of time, effort and information from the following "all star" cast:

Cheryl Blake, who handled all the secretarial work and the many related tasks of manuscript preparation. Adrianna Ortisi who arranged publication with the printer and supervised production and graphic arts, and the Editorial Assistants who did the proofreading. Robert W. Grant who loaned books and 19th century newspapers from his private collection, and Anthony R. O'Neill who also provided reference books. George H. Tryon for his editorial contribution to the aircraft fire summary. Arthur Washburn who searched through hundreds of fire reports for the pictures and information which are such a vital part of this book. The editors, librarians and other members of the staff of the National Fire Protection Association and Warren Y. Kimball and Percy Bugbee who offered advice and assistance in the planning and preparation of the text.

—— and these chiefs and fire officers who provided historical information on their fire departments ——

Chief John L. Swindle, Birmingham, Alabama Fire Department; Chief Gene Bennett, Anchorage, Alaska Fire Department; David L. Roth, Phoenix, Arizona Fire Department; Chief Jack D. Davis, Little Rock, Arkansas Fire Department; Chief Engineer Kenneth R. Long and Captain Silas Clarke, Los Angeles, California Fire Department; Chief Keith P. Calden, San Francisco, California Fire Department; Chief Francis J. Sweeney, New Haven, Connecticut Fire Department; Chief James P. Blackburn, Wilmington, Delaware Fire Department; Battalion Chief Dennis N. Logan, Washington, DC Fire Department.

Chief D. A. Hickman, Miami, Florida Fire Department; Chief John M. Schroder, Savannah, Georgia Fire Department; Chief of Department, Honolulu, Hawaii Fire Department; Deputy Chief Fire Marshal J. T. Deichman and Robert A. Freeman, Chicago, Illinois Fire Department; Deputy Chief Donald E. Bollinger, Indianapolis, Indiana Fire Department; Acting Chief Thomas O. Munley, Louisville, Kentucky Fire Department; Lieutenant Richard K. Blackmon, New Orleans, Louisiana Fire Department; State Fire Marshal James C. Robertson, Baltimore, Maryland; Chief Thomas J. Burke, Baltimore, Maryland Fire Department; Clarence C. Woodward, Historian, Detroit, Michigan Fire Department.

Chief Charles Kamprad, St. Louis, Missouri Fire Department; Special Projects Chief Don J. Dougherty, Omaha, Nebraska Fire Department; Liaison Officer Larry P. Powell, Sr., Las Vegas, Nevada Fire Department; Deputy Chief Joseph McLaughlin, Newark, New Jersey Fire Department; Chief John Lee, Charlotte, North Carolina Fire Department; Chief R. D. Shipp, Wilmington, North Carolina Fire Department; Chief W. T. Stewart, Fargo, North Dakota Fire Department; Chief Bert A. Lugannani, Cincinnati, Ohio Fire Department; Chief Byron D. Hollander, Oklahoma City, Oklahoma Fire Department; Chief Gordon A. Marterud, Portland, Oregon Fire Department.

Fire Commissioner Joseph R. Rizzo, Philadelphia, Pennsylvania Fire Department; Assistant Chief Richard Rebello, Providence, Rhode Island Fire Department; Chief Guthke, Charleston, South Carolina Fire Department; Chief W. J. Tellinghuisen, Sioux Falls, South Dakota Fire Department; Chief Merrell C. Hendrix, Dallas, Texas Fire Department; Chief I. O. Martinez, San Antonio, Texas Fire Department; Chief Leon DeKorver, Salt Lake City, Utah Fire Department; Chief J. P. Finnegan, Jr., Richmond, Virginia Fire Department; Lieutenant S. E. Gibson, Training Division, Sandston, Virginia Fire Department; and Chief William Stamm, Milwaukee, Wisconsin Fire Department.

—— and these individuals in the fire protection industry who provided pictures and helpful data ——

Henry Kalser, American LaFrance, Elmira, New York; P. A. Coombs, Cairns and Brother, Inc., Allwood, Clifton, New Jersey; Charles R. James, Howe Fire Apparatus Company, Inc., Anderson, Indiana; G. Bowman, Hale Fire Pump Company, Conshohocken, Pennsylvania; Bert Nelson, Mack Trucks, Inc., Allentown, Pennsylvania; George Layden, Peter Pirsch and Sons Company, Kenosha, Wisconsin; H. Roth, Seagrave Fire Apparatus, Inc., Clintonville, Wisconsin; Peter Mahoney, G & W Eagle Signal, Davenport, Iowa.

Introduction

In the midst of the continual movement which agitates a democratic community, the tie which units one generation to another is relaxed or broken; every man readily loses the traces of the ideas of his forefathers, or takes no care about them.

From *Democracy In America* by Alexis de Tocqueville — 1848

During the past three centuries the people of the United States have been exposed to millions of dangerous and tragic fire situations, sufficient to give them more experience with fire than any other nation in history. America was susceptible to fire long before its independence was achieved in 1776. Fire threatened to wipe out the first colonies. Fire was used extensively as a savage weapon of intimidation and murder during the French and Indian Wars. Fire became a greater problem as seaports and other large, expanding communities suffered tremendous, sweeping conflagrations, and terrible loss of life in single building fires and explosions. Fire was used as a tactical and strategic weapon in war.

Fire devastated the crops, the farms, the forest lands. Fire and explosions have been used as a vicious means of social and political protest.

Fire has destroyed uncounted thousands of human lives, and has injured or otherwise affected millions of people during the two centuries in which the United States of America has served as a self-governing nation.

This book presents an illustrated narrative of the history of fire in the United States. It is only a summary; a condensed representation of many incidents, of the growth of fire departments, and of our combined efforts to resist and control this repeated threat to our safety.

In such historical review, we can identify the critical periods of change, the trends that indicated improvements in our knowledge, our technical advances, and our solutions of the fire problem. Here are some of these:

The first legislation resulting from civic response to fires in colonial times; the first volunteer fire fighting action with buckets and hooks; the beginning of fire patrols and mutual aid societies; the laws establishing the first public fire departments; and finally, the establishment of paid fire departments with authority to inspect properties and otherwise extend fire prevention.

Improvement in the means of applying water on the fireground, from hand buckets, to hand pumps, to horse-drawn pumps, to steam-powered pumps, to self-propelled apparatus, to modern centrifugal pumps, to aboveground application by aircraft.

Improvement of water systems — from wells, to cisterns, to wooden piping, to metal and concrete asbestos underground water mains, to the complex, highly controlled pumping and hydrant systems that serve our modern cities and towns.

Change in the kinds of materials used for buildings, from the grass and thatched roof dwelling; to mud or clay covering; to brick and cement; to reinforced concrete; to our present fire-resistive structures having modern fire detection, alarm and extinguishing systems.

Change in the use of power for energy — from the wood fire; to coal; to steam; to dammed rivers and lakes; to wind powered machines; to oil; to electricity; to nuclear generating plants; to hydrothermal and solar energy.

Development of extinguishing agents, from dirt to water; to water additives; to other liquids and chemicals; to gases; to portable extinguishers; to wheeled extinguishers; to fixed extinguishing systems; to aircraft dropping extinguishing agents on ground fires.

Development of our means of analyzing fire causes, of educating the public and specific groups who can take responsible actions; the statutory authority of municipal fire departments; the establishment of fire protection associations; and national standards of fire protection.

Surely, from what we have learned, the control of fire is within our collective grasp. Surely, when we consider the many problems of protecting ourselves against fire, and when we review what tragedies and devastation have accompanied fire in past years, we should be able to avoid the regression so cynically predicted by Alexis de Tocqueville. Surely, we must remember, and act upon, the lessons of past experience with fire.

Frontispiece — In 1775 British ships sailed into Falmouth (now Portland, Maine) to burn that community as punishment for aggressive actions of the colonists. The fleet bombed and burned 139 houses, 278 stores and other buildings but the people of Falmouth remained and eventually repulsed the enemy.

1600-1800

Chapter I

"It is ordered that fower good strong Iron Crooks with Chaines and rops fitted to them, and thes Crooks fastned on a good strong pole be forthwith provided by the seleckt men, which shall hang at the syd of the meeting house, thear to be ready in Case of fier."[1]

One fact is clear: if the first colonies that settled on the eastern coast of America had been devastated and destroyed immediately, they would have been replaced quickly by new settlements. The governments of England, France, Spain, The Netherlands and the Norwegian countries were preparing hastily to occupy the New World. Many thousands of people in these countries were ready to escape from their homes and traditional environments to seek a new life of freedom and promise in the American colonies. Even prospects of a very dangerous ocean voyage, and the threat of famine, disease, Indian attack and other disasters in the new land could not hold back the increasing masses of these new settlers.

But those first colonies were almost wiped out! Harsh weather, poor crops, resistance of local Indian tribes and the general confusion of settling in the new land severely tested the will and survival capabilities of the new settlers.

From the beginning, the most consistent, terrifying, repeated experience was fire. This should have been expected at the time, for there were no organized fire control forces, no plans or legislation to minimize the hazards of fire, no readily available water supply for fire fighting, and the means of communication were minimal. The colonists probably had vivid recollection of previous fire disasters in their European homelands. In this new country shelter for housing and storage of food and other valuables was created by the most

expedient means. Some of the colonists moved into natural caves, or dug holes in the sides of hills, trying to create temporary, livable shelters that would offer protection against the harsh elements of winter. Others gathered brush, marsh grasses and clay to build fragile, thatched-roof dwellings, hardly larger than small cottages. Within these buildings, the central fireplace was made of bricks and mud or clay. The chimney, placed to vent smoke and heat, was made of "daub and wattle," brush and stalks or wood planks, covered by a heavy mixture of mud and clay. Open cooking utensils of earthenware or metal were in constant use, sometimes filled with wax or other flammables. Beer, whiskey, brandy and gunpowder were natural provisions in these early homes, and subsequently were to be important factors in the sweeping fires that devastated the larger communities. In those first settlements, the small dwellings, barns and storage buildings were built close together, usually within the protection of a nearby fort that served as defense against attack by Indians. Such close grouping helped most fires to move quickly from building to building.

It did not take long for the settlers to learn about fire. Jamestown, Virginia, the first colony to be settled (1606) under Captain John Smith had a devastating fire in 1608 that destroyed most of the colonists' lodgings and provisions.[2]

In 1613, Dutch sailing vessels landed on Man-

1

Letter from Plimoth Plantation[4]

This fire was occasioned by some of ye sea-men that were roystering in a house wher it first begane, makeing a great fire in very could weather, which broke out of ye chimney into ye thatch, and burnte downe 3. or 4. houses, and consumed all ye goods & provissions in ym. The house in which it begane was right against their storehouse, which they had much adoe to save, in which were their comone store & all their provissions; ye which if it had been lost, ye plantation had been over-throwne. But through Gods mercie it was saved by ye great dilligence of ye people, & care of ye Govr & some aboute him. Some would have had ye goods throwne out; but if they had, ther would much have been stolne by the rude company yt belonged to these 2. ships, which were allmost all ashore. But a trusty company was plased within, as well as those that with wet-cloaths & other means kept of yey fire without, that if necessitie required they might have them out with all speed. For yey suspected some malicious dealling, if not plaine treacherie, and whether it was only suspision or no, God knows; but this is certaine, that when ye tumulte was greatest, ther was a voyce heard (but from whom it was not knowne) that bid them looke well aboute them, for all were not freinds yt were near them. And shortly after, when the vemencie of ye fire was over, smoke was seen to arise within a shed yt was joynd to ye end of ye storehouse, which was watled up with bowes, in ye withered leaves wherof ye fire was kindled, which some, runing to quench, found a longe firebrand of an ell longe, lying under ye wale on ye inside, which could not possibly come their by cassualtie, but must be laid ther by some hand, in ye judgmente of all that saw it. But God kept them from this deanger, what ever was in-tended.

Governor William Bradford

hattan Island to conduct trade with the Indians near the mouth of the Hudson River. These ships were named the *Tiger* and the *Fortune.* Shortly after the traders had constructed a few wooden dwellings, the *Fortune* returned to Holland but the remaining ship, the *Tiger*, burned completely, and the small group of Dutch settlers remained isolated for a year until they could construct a new ship. In 1615 the *Fortune* returned with other ships and the settlement of Nieuw Amsterdam became estab-lished by the Dutch West Indies Company.[3]

The colony at Plimouth (Plymouth), Massachu-setts, settled in 1620, three years later had a fire sweep through the town destroying nearly all the provisions and at least seven dwellings. The first Puritan Governor, William Bradford, carefully noted the occasion in his records, blaming some roistering sailors as the cause.[4]

The first fire recorded on the island that was to become Manhattan occurred in 1628 when a dwell-ing caught fire and burned to the ground. Subse-quently, during the reign of Governor Peter Stuyvesant, the first fire prevention and building code was established. Among other things, it allowed for creation of a committee to inspect all chimneys in Nieuw Amsterdam with the privilege of fining owners who did not keep chimneys swept clean. Stuyvesant also began a form of municipal fire protection, purchasing buckets for use during fires and establishing a fire warning system that featured selected individuals as fire watchers.[5] These persons carried wooden rattles that made noise loud enough to rally nearby citizens for fire fighting.

In 1630, the town of Boston was considered the wealthiest and most populous in Massachusetts Bay colony. It had some mud houses but these were few and were occupied by the very poorest of colonists. Most buildings were of wood although some had been constructed of stone and brick, but until the town was at least twenty years old these were exceptions. The first dwellings were generally one story in height, with roofs covered with thatch or boughs of trees. As time passed, those who could afford the expense built their houses with two stories in front with a shingled roof that ran nearly to the ground in the rear, leaving but one story exposed in back. Later, lapped or double roofs were used on these buildings.[6]

No serious fires were recorded in Boston until March 16, 1631, when Massachusetts Governor Winthrop noted the following:

About noon the chimney of Mr. Thomas Sharp's house in Boston, took fire, the splinters being not clayed at the top, and taking the thatch, burnt it down. The wind being north-west, drove the fire to Mr. Coulburn's house being [some] rods off, and burnt that down also which were as good, and as well fin-ished, as the most on the plantation.

Much of their household stuff, apparell and other things, as alsoe some goods of others who sojourned with them in their house, were consumed; God so pleasing to exercise us with corrections of this kind as he hath done with others; for the prevention whereof in our new towne intended this somer to bee builded, we have ordered that noe man there shall build his chimny with wood, nor cover his house with thatch, which was readily assented onto; for that divers others houses have beene burned since our arrival (the fire alwaies beginninge in the wooden chimny), and some English wigwams, which have taken fire in the roofes covered with thatch or boughs.

The first charter for the Jamestown colony of Virginia was issued in 1606, followed by a second charter three years later, and a third in 1612. Massachusetts was started with the Mayflower contract in 1620, but subsequently the colonies which were to become New Hampshire (1623), Rhode Island (1636), and Connecticut (1639) separated from Massachusetts but later joined in the New England Confederation in 1643.[7]

New Jersey was separated from Nieuw Amsterdam when the British took that island from the Dutch in 1664 and the former Dutch settlement was named New York.

Delaware (1634), North Carolina (1660), South Carolina (1670), and Georgia (1733) formed the remainder of the thirteen original colonies. Most of these settlements had severe fire experience in those first hundred years, formed volunteer fire departments and took similar measures for fire control. Unfortunately, many records and reports of those early communities have disappeared.

Swab used by householders and fire wardens for smothering fires in thatched roofs.

The founding of New York, New Jersey and Delaware resulted after a war between the British and the Dutch. In 1644, James, Duke of York, England, conquered Nieuw Amsterdam with a British fleet and became proprietor of the colony. It became a royal colony when he became King James II of England, then was named New York.

New Jersey was part of Nieuw Amsterdam and James gave this colony to two friends as proprietors. At one time it was divided into East and West Jersey.

Delaware also was partitioned after the same war. Originally, it had been founded by a Swedish group headed by Peter Minuit, but it became a Dutch colony, then a proprietory English colony in 1664.

"He has plundered our seas, ravaged our Coasts, burnt our townes, and destroyed the lives of our people . . ."

From the Declaration of Independence

1653

The select men have power and liberty hereby to agree with Joseph Jynks for Ingins to Carry water in Case of fire, if they see Cause soe to doe.

1654

It is ordered that iff anye Chimney shall be fired soe as to Flame out att the topp the owner thereof shall paye unto the towns treasurer, for the use of the towne, the sum of five shillings, and the order made about Chimneys the 24th: 9th: 51, is hereby repealed.

1654

Itt is ordered that noe man shall take and Carrye away anie of the towns ladders from eather of the meeting howses, except in Case of fire; in penaltye of five shillings for every ladder Carryed away Contrary to this order.

Type of hook used for pulling down houses when fire was spreading in colonial settlements.

The Boston conflagration of 1653 probably was the first for which fire deaths were recorded. Governor Winthrop's report stated that three children of a Mr. Sheath were victims of this blaze in the center of what is now downtown Boston.[8] The citizens were greatly concerned about saving the town against the destruction of fire and at a town meeting in March that year they passed regulations specifying the use of ladders, chains, ropes and other fire control equipment. The records of that town meeting included the first mention of storing water for fire protection, as follows:

William Franklin and Neyghbors about his howse is granted liberty to make a cistern of 12 foot or greater, if they see cause, at the Pumpe which standeth in the hie way near to the State arms Tavern (*corner of present State and Exchange Streets — Ed.*), for to howld watter for to be helpful in case of fier, unto the towne. He is to make it safe from any danger of children.

Arson was a serious and frequent problem all through the 17th, 18th and 19th centuries as indicated by this portion of the law passed by the General Court of Massachusetts in 1652:

Enected by the Authority of this Court that any person or persons whatsoever of the age of Sixteen years and upward, that shall after the publication hereof wittingly and willingly set fire any Barn, Stable Mill, out House stack of wood, Corn or Hay, or any other thing of like nature, shall upon due conviction by testimony or confession, pay double damage to the party damnified, and be severely whipt.

Typical fireground action in 1776, with handtub in the foreground, bucket line at left, and nozzle on leather hose at right. By this time, dwellings had clapboard siding, wood shingle roofs and many other combustibles. (Bettman Archives.)

Another portion of that law was more severe. It specified that any person legally convicted "by due proof, or confession of the Crime . . ." (of burning a building) could be sentenced to death and forfeit "much of his lands, goods, chattels necessary to make full satisfaction to the party or parties damaged by the fire."[9]

The History of the Boston Fire Department had these comments on the arson problem:

About this time many daring attempts were made by incendiaries to destroy the town, and, judging from the methods adopted, they were the studied plans of some secret and determined gang of "fire-bugs."

On January 9, 1677, John Hull records in his Diary that —

A candle was fastened to the roof of a house and burnt through the roof, yet was prevented spreading through the wonderful Providence of God, but the author not known. A barn of Mr. Usher's was burnt about 1 o'clock in the morning of July 5, although his house and other property were saved. August 6. Be-

tween two small houses of Mr. Bradons, situated in Shrimpton's Lane (Exchange street), was discovered, about 10 o'clock at night, a lighted candle. About one hour later an attempt was made to set fire to a barn in Usher's Lane (near Atlantic avenue and Bedford street), but the hay, being cut on the salt marsh, smothered, and was discovered in time to prevent it bursting into a blaze.

Several other attempts were made in different sections, but were unsuccessful.

Regarding these incendiaries the General Court, on October 10, passed the following act: —

Whereas many secret attempts have been made by evil-minded persons to set fire to the town of Boston and other places tending to the destruction and devastation of the whole; this Court doth account it their duty to use all lawful means to discover such persons, and prevent the like for time to come.

It was then ordered that all persons, inhabitants or strangers, should take oath of allegiance, which oath was vastly different then to that prescribed in 1652. For this end constables and tithing men were to make a canvas of the town every three months.[10]

In 1785 this was one of two pumping engines used in Old Salem (Winston-Salem) North Carolina. Two men would work the "brakes" at each end, while another directed the gooseneck nozzle in top center. Note leather bucket, wooden tank and wooden wheels. Tricorner hat, jackets and vests, knickers and stockings and buckle shoes were typical of clothes in that period.

1657

Mr. Tho. Broughton is fined five shillings for his chimney being on fire and flaming out.

Ben Gillam is fined ten shillings for making a fire upon the wharfe.

1658

Itt is ordered that Jno. Marshall goe through the towne to view whether every house is supplied with a ladder according to order of towne, and to returne the names of such as are defective.

1658

Whereas many careless persons carry fire from one house unto another in open fire pans or brands ends, by reason of which greatt damage may accrew to the towne, It is therefore ordered that no person shall have liberty to secyre itt from the wind, upon the poenalty of ten shillings to bee paid by every party so fetching, and halfe so much by those that permitt them so to take fire.

1659

Itt is ordered that in case of fire breaking outt in any parts of this towne, which may possibly threaten ruine to a greatt part thereof iff not seasonably prevented by pulling downe some house or houses to that end; Itt shall be lawful for the major part of the magistrates, Comissioners, and select men, that shall then bee present att the fire, or for any three of them mett together, and two of them concurring, to cause any house, or part thereof, to bee, puld downe; And that whatever house or part thereof bee puld downe by their order, shall again bee repayred and made good by the towne to him or them who shall so have their houses puld down or impayred; and the former order in p. 105 of this booke is hereby repealed.

1659

Itt is ordered that no person, whether watchman or any other, shall, att any time, take tobacco, or bring lighted match, or fire underneath or aboutt any part of the towne house except in case of military exercise, upon the penalty of twenty shillings for every such offence, except under covert for use of the house above.

In the Session of the General Assembly in Maryland, February 23 through March 19, 1638, the first law relating to arson was enacted. This decreed: "Malicious trespasses as to burn or destroy wilfully a house or stack of corns or tabacco" as a felony. Penalties established required the offender to "suffer the pain of death by hanging"; however, a less severe penalty was permitted: the offender could "lose his hand or be burned in the hand or forehead with a hot iron and shall forfeit all his land." A second offense required the death penalty.

John Anderton of Calvert County was the first person to be charged under the above law. He was charged with setting fire to a tobacco curing house which contained 12,000 pounds of tobacco. The building burned to the ground.

A "non-suit" was granted as a result of the trial in that evidence was presented that the defendant had partial ownership of the property involved and was therefore not considered a trespasser.[11]

Colonial dwellings and business buildings were constructed of rather flimsy material; therefore, when fire occurred, it was quite simple to pull down these buildings to create separation and minimize fire spread.

Most of the colonial towns and settlements had bucket brigades which could feed the hand-powered pumps, or might wet down buildings in advance of a moving fire. However, in the absence of heavy streams from handlines or force pumps, the best alternative was to create separation between buildings. Usually, iron hooks attached to ropes were sufficient for this purpose, since they could quickly pull thatched roof from a dwelling and, if necessary, tear down the walls. Later, when more substantial buildings were constructed of clay and brick, explosives were used to achieve the same sort of fire break. This sort of action was particularly important in the time before more powerful hand pumps were developed.

Because such buildings had to be destroyed for the general safety of the community, in March, 1658, a law was passed in Massachusetts to compensate owners of buildings destroyed in this manner to halt the advance of fire.

Boston had two other severe conflagrations: one in 1676 and one in 1679. Each destroyed a major portion of the built-up center of town. One of the biggest obstacles and hazards was the presence of gunpowder in almost every building. The colonists used gunpowder for their muskets as well as for blasting operations on their farms, so it was common practice to store barrels of this explosive in a convenient place within the building. In one early fire, eighteen barrels of gunpowder exploded, destroying several buildings and many important city records. Subsequently, in the 1653 fire, Bostonians saw buildings exploding one after another as the fire involved gunpowder in each dwelling.

In November, 1676, another major fire occurred in Boston, destroying forty-six dwelling houses, several warehouses and stores and the North Meeting house. The fire spread so rapidly that all these buildings were devastated within four hours. Fortunately, a rainstorm with strong winds helped extinguish this fire which could have destroyed a much larger portion of the town. This seems to have been the first time when explosives were used deliberately to destroy buildings and hinder the spread of the fire front, a practice that was continued into the twentieth century. Considering the flammability of colonial buildings, it probably was simple and effective to blow them up quickly, but in later conflagrations, when buildings were larger and more numerous, such destruction seemed to be ineffective as a means for stopping the spread of fire or isolating its position.

A second group of fires started by arson occurred during the spring and summer of 1679. Seventy warehouses and eighty Boston dwellings were destroyed on the night of August 8 and trading vessels lying in the harbor were also burned. As a consequence of the rapid spread of the fire among buildings, the General Court, immediately after the fire, established the first legislation to control the construction of buildings, as follows:

The Court having a sense of the great ruin in Boston by fire and hazzard still of the same by reason of joining and nearness of their buildings, for prevention of damages and loss thereby for future. Do therefor order and enact that henceforth no dwelling-house in Boston shall be erected and set up except of stone or brick and covered with slate or Tyle, on penalty of forfeiting double the value of such buildings, unless by allowance and liberty obtained from the Magistrate, Commissioners and Selectmen of Boston or major part of them, and further the Selectmen of Boston are hereby empowered to hear and determine all controversies about properties and rights of any person to build on the land wherein now lately the housings hath been burnt down, allowing librty of appeal for any person grieved, to the County Court.

The Boston report comments wistfully: "Had this law continued in force we would not in all probability have such a large fire record for our city."[12]

Charleston, South Carolina, was another of the early communities to experience fire disasters. In 1700 and again in 1740 the settlement was practically wiped out by fire. Then, during the Revolutionary War, on January 15, 1778, half the city went up in flames including the original Huguenot

In 1774 George Washington was a member of the Friendship Fire Company, Alexandria, Virginia. In that year, while attending the Continental Congress in Philadelphia he purchased this fire engine for eighty pounds, ten shillings — equal to $400. Here the engine is in front of the building housing the present Friendship Veterans Fire Engine Company in Alexandria which celebrated its 200th Anniversary February 18, 1974.

Church, 250 dwellings, stores and warehouses, and the Charleston Library Society, which contained nearly 6,000 precious volumes of literature. Later in 1779 and in 1861 the city suffered other great fires.[13]

As a consequence of these calamities the people of Charleston organized to develop fire control capabilities. In 1784, a "hand-in-hand" company was formed "to cope with the growing menace of fire." This group's operation consisted of passing buckets from water cart to the fire, an awkward and ineffective means of controlling any blaze except a small one; but this is believed to be the start of the volunteer fire company in Charleston.

About the same time, the City of Richmond, Virginia, was having severe fires and was beginning

"One fire was started in 1800 when a citizen investigated a leak in a barrel of brandy. He was using a candle to see what the trouble was. The brandy ignited and the keg of gunpowder nearby added to the merriment of the occasion. In modern times, the man who looks for a gas leak with a lighted match is no improvement." *From the History of the Charleston, South Carolina Fire Department.*

Two hand buckets in use at the time of the Revolutionary War. (Smithsonian Institute.)

initial fire fighting organization. In 1787, a blaze swept through a large portion of the city destroying almost fifty buildings with property damage estimated at nearly half a million dollars. The people of Richmond had only private wells and springs for water supplies but some public wells were located at street corners in the city. At the sound of an alarm, people would respond carrying family pails, then form a line to pass pails of water from the nearest wells to the fire. This bucket brigade, like Charleston's and others, was relied upon for some years.

Near the end of the century fire insurance came to Richmond in the form of the Mutual Assurance Society. According to a historical report issued by the Richmond Bureau of Fire, the first insurance policy in the United States was written on the Masonic Hall, which is still on East Franklin Street in that city, and the policy is still in effect.[14]

In Providence, Rhode Island, until 1754, there seemed to be little action by the government of the colony, or the town, to protect property or life from fire. In that year, however, the inhabitants of the "compact" part of Providence petitioned for power to purchase a "large water engine." A law was also passed requiring each housekeeper to be provided with two fire buckets, of leather, two-gallon size. These were supposed to be kept in the front part of the house and were to be used only in case of fire. The law required that the owner's name be painted on each bucket, and a five dollar fine was specified for violation of this rule. Records do not indicate when the first engine was delivered, but in 1760, a second engine was purchased from Boston and engine men were appointed in June 1763.

Providence, at yearly town meetings, chose three gentlemen "endowed with sound judgment, fidelity and impartiality" as Presidents of the Fire Wards. Their duty was to superintend the use of gunpowder for blowing up buildings to check a conflagration. The owners of these buildings later were compensated by the town for all damages, unless the fire started on the premises of their buildings, in which case "all was a loss to the owners."

The law also provided for appointment of persons to act as Fire Wards. The badge of office provided for these individuals consisted of a red and white trumpet. In case of fire they were supposed to take this to the fireground, vigorously exert their authority and inflict fines of four dollars for any disobedience of their orders. The income obtained was distributed among the people "most distressed by the fire."[15]

By 1792 Providence had four engines of the Newsham type. (See page 13.) "Forming a line" of bucket passing was the primary duty of the Fire Wards.

Other colonial towns that would grow to modern cities began their efforts for municipal fire protection in very simple ways.

In 1948 the United States Post Office issued a three cent stamp honoring the 300th Anniversary of volunteer firemen. Peter Stuyvesant, Governor of Nieuw Amsterdam (New York) was named organizer of the first volunteer firemen in America.

Fire Department Growth

The appointment of four fire wardens by Nieuw Amsterdam Governor Peter Stuyvesant in 1648 introduced a new measure of municipal fire protection, the inspection of properties. These wardens had the power to inspect chimneys in the Nieuw Amsterdam buildings and to fine building owners a sum of four gilders for each chimney that was not swept properly. (Chimney sweeping was a common practice in European countries.) This money became part of a fund for the purchase of buckets, hooks, and ladders to be used in fire fighting. Unfortunately, when the fire wardens and Governor Stuyvesant assumed too much authority, they lost the trust of the people. One source mentioned that, in 1664, when the British were anchoring in the harbor during a war with the Dutch, Stuyvesant tried to gain the support of the citizens but the people did not resist the British who easily took over the colony and named it New York.[16]

Prior to that incident, in 1659, Governor Stuyvesant had appointed a "Rattle Watch," selected individuals who patroled the streets from nine at night until the beating of a drum signaling the arrival of dawn. Essentially, this was the first municipal fire alarm system.

Ten years later, in 1669, a "Brent Meister" was appointed to be in charge of the inspectors and wardens.

It was in this decade that a change occurred in England which was to influence the growth of American fire departments. After the great fire of London in 1666, there was no public fire brigade but the fire insurance business rapidly became established. These companies, in order to protect their insured properties, set up private fire brigades, equipping them with an engine, thirty buckets, two squirts and three ladders. These brigades developed serious rivalries with each other to secure new insurance clients by competing in the efficiency of fire protection provided.[17] It was not long before similar violent competition developed in America when volunteers assumed fire fighting responsibilities.

Many historical publications have reported the work of Benjamin Franklin who was given credit for establishing the first volunteer fire department in Philadelphia. After living his early years in Boston, he became publisher of the *Pennsylvania Gazette* and was able to urge new developments for the city. First came the organization of a fire watch system and next the formation of the Union Fire Company on December 7, 1736. This group had thirty members, one hand-pulled pump, 250 buckets, thirteen ladders, two hooks, and one 8-foot "hause" of rope. Simultaneously, a number of insurance companies were formed and they in turn sponsored other fire companies.

Savannah, Georgia, purchased a hand engine in 1759 and fifteen townsmen organized a fire company, agreeing "to keep the engine in good repair and attend upon any accident of fire." In the same year, wardens and vestrymen of Christ Church in the town were authorized to procure fifty leather buckets and fifteen fire hooks, to be paid for by a "tax upon the citizens in proportion to the number of hearths in house." Thus was the first fire company and first ordinance formed in Savannah.[18]

In Baltimore, Maryland, on September 22, 1763, a number of settlers met and formed the Mechanical Company. For many years the members of this volunteer company discharged nearly all the duties needed for the town government, including fire fighting, police work and court functions. When more volunteer units were formed, the Mechanical Company was renamed Mechanical Fire Company No. 1. This lasted until December 7, 1858, when the city council passed an ordinance establishing a paid fire department. Then, on February 15, 1859, the new paid fire department began. It consisted of a chief engineer, two assistant chief engineers, one fire inspector, six engine companies and one hook and ladder company.

One interesting event occurred in March 1849 when the Mechanical Fire Company's suction engine "Comet" was shipped aboard the clipper ship *John Marshall* which sailed from Baltimore Harbor

A "Garden Type Fire Engine" with a hand pump, suction chamber, and short length of hose with nozzle. This was used for small fires in early 18th century. (Photo by Franklin Institute.)

"Old Shag Rag" one of three engines imported from England by Ben Franklin when he founded a fire company in Philadelphia in 1764. It was delivered to a volunteer fire company known as the "Gentlemen of the Middle Ward." (Photo by Franklin Institute.)

bound for San Francisco. The engine had been purchased by former Baltimore citizens residing in San Francisco. In that period of the gold rush they wanted to organize a fire company "similar to the Mechanical."[19]

In Wilmington, Delaware, on December 22, 1775, citizens organized Friendship Fire Company No. 1 to protect the Borough of Wilmington from fire. Each member pledged to furnish one large wicker basket and two small leather buckets for fire fighting. For ready identification at fires, each member also wore a red leather chest protector and a high red hat. At the end of December the Borough Council of Wilmington purchased two Newsham engines from a French man-of-war lying at the Port of Wilmington. (*See "alarms," page 68.*)[20]

The Newark, New Jersey Fire Department dates back to January 1797 when a meeting was held to confirm the purchase of a fire engine and formation of two volunteer fire companies. Instead, only one fire company was organized and a small hand engine was purchased. Not long before this "the Patriotic Society for Promoting Objects of Public Utility" had purchased a number of ladders "to be kept sacred for the use of citizens in case of fire." Since 1668, there had been a regulation requiring all householders to provide ladders for such use but it was mostly ignored. The burning of the "Boudinot" house in the town before 1797 shocked people into action and a "Fire Association" was formed. Sub-

scriptions from one hundred leading citizens provided funds to buy hooks, ladders, two little fire engines and a leather bucket for each member, "to be hung with his hat in his front hall."[21]

In 1798 the General Assembly of Kentucky enacted a law allowing formation of five fire companies in Louisville. This required that names of all members be recorded in county court records and restricted the volunteer membership to forty men per company. From this date until 1832 the city was protected by the volunteer fire brigade which included active, athletic men, well-trained and operating under strict discipline. General fireground practice was to form a double row of these members to pass buckets of water to and from the fire. In severe fire circumstances sometimes women were pressed into this type of service.[22]

Even before the Revolutionary War was drawing to a close, people in the eastern colonies were already moving to the new lands of the West and existing communities were changing. Kentucky became a state in 1792, Tennessee in 1796, and down in Maryland, where the town named Georgetown was created by the State Assembly in 1751, the town population had grown to more than 1,500. By 1789 this town had collected enough money to buy a hand engine and fire buckets and every male inhabitant old enough to vote became a member of this community's first fire department. Just three years later, in 1792, the first executive mansion of the new nation was built in Washington.[23]

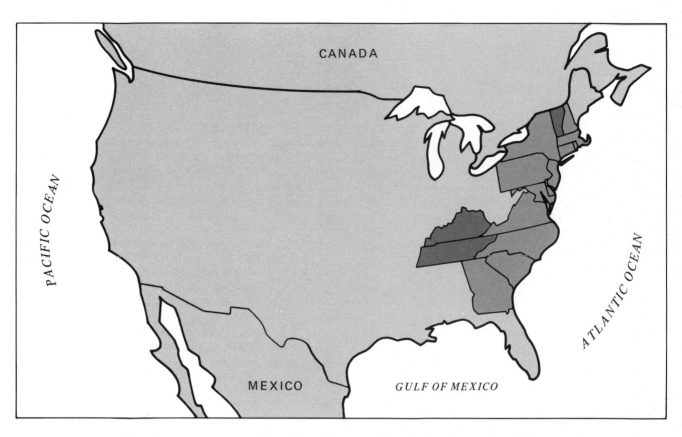

Kentucky (1792) and Tennessee (1796) became states just before the start of the 19th century.

Rescue of a child.

This old Dutch print of a 1669 fire shows fire fighting methods of that time. Note use of squirt (hand pump and gooseneck jet at left). A large number of persons and the general confusion at such fires was frequently represented in paintings, engravings, and prints of colonial times.

Developments in Fire Apparatus

The use of fire fighting equipment and the methods of fighting fires in American colonies were greatly influenced by fire control practices in Europe. The devastating conflagrations in European countries during the 16th and 17th centuries had led to some practical fire control measures, but not much attention had been given to the means of distributing water for fire fighting. Even though pumps and water systems had been used back in the days of the Egyptian dynasties and the Roman Empire, these practical measures seemed to have been forgotten until the beginning of the 17th century.[24]

An illustration of a fire pump described in 1612 by Heinrich Zeising showed a two-cylinder force pump with a swiveling delivery nozzle, mounted in a tank on wheels. Later, an air tank was added to maintain pressure in the delivery side of the pump so that water could be discharged in a continuous stream. This invention was attributed to a Mr. Hautsch of Nuremberg, Germany, but was not adapted to English pumps for another half century. In the meantime, hand pumps in the form of a syringe, a fire squirt, or a small stirrup pump with attached hose, were kept available for controlling small fires.

In the American colonies the first mention of a water engine for fire fighting appeared in a record of the town of Boston for March 1, 1653–4.[23] This stated that "the select have power and liberty hereby to agree with Joseph Jynks for Injins to convey water in case of fire, if they see cause so to do." Later, there was no record that a contract was ever made with Joseph Jynks, nor was such equipment delivered. Some historians have concluded that Jynks only made a syringe or some other small hand pump, identified in those days as an "Ingine". However, the growing town, already deeply concerned about fire protection, faced the problem of making water available.

In 1649, a William Myng arranged to develop wells on his estate and to place wooden pipe for establishing a conduit and water works. Three years later the General Court granted an act of incorporation to the inhabitants on Conduit Street in Boston (*near the present Government Center — Ed.*) to provide a fresh supply of water to their families and especially for use in case of fire. This reservoir was about twelve feet square, made for holding water conveyed to it by wooden pipes leading from neighboring wells and springs. Over the reservoir was a wooden building used for storage; later this was removed and the conduit was covered by planks raised in the center about two feet sloping to the sides. Although this conduit subsequently fell into disuse and became a public nuisance, the selectmen and citizens of the town were becoming much more aware of the importance of water for fire protection.[25]

Two other inventions in Europe were soon to increase capabilities for fire fighting. In Amsterdam, Holland, the Superintendent of the Fire Brigade, Van der Heiden, introduced the use of fire hose made of leather to be attached to the gooseneck delivery nozzles of force pumps then being used in 1673. The tank of the pump was connected by flexible hose with a canvas trough in a wooden stand, placed on a canal bank and kept filled with water from buckets. Van der Heiden subsequently designed wired suction hose to be used with this pump. In addition to his fire brigade duties he was a distinguished artist and author of a Dutch book on fire protection entitled "Slangbrand-Spuiten."[26]

Another Dutch print shows fire involving the Amsterdam rope works in 1673. According to Van der Heiden the goose neck pumps were used for the last time at this fire. New leather hose replaced them. (Fire Engines and Other Fire Fighting Appliances.)

The American Museum of Fire Fighting in Hudson, New York (center spread) has a Newsham hand engine that was imported from London in 1731. This is a reproduction from the museum's brochure. It mentions that a Newsham of the "sixth size" had a cistern of 170 gallons capacity and a body length of 6 feet, 8 inches, 22½ inches width, and a deck 25 inches high. The engine was not capable of suction but had to be filled by hand buckets. (American Museum of Fire Fighting.)

It was not until 1721 and 1725, when Richard Newsham of London, England, patented a new kind of pump that a remarkable change in fire fighting methods became evident. The New York Fire Department obtained two of these pumps in 1734. Each machine had two single acting pumps and an air vessel placed in a tank which formed the frame of the machine. The pumps were worked by men pushing on long cross handles which projected at the side of the pump, but there were two treadle boards near the center of the machine where additional men could assist the pumping by exerting their weight on the descending treadle. The old style swivel nozzle could be used, but the Newsham also was designed to supply riveted leather hose at the end of which was a long, small-tipped nozzle. These engines were mounted on solid wooden block wheels and were pulled by the fire fighters. Because of their size and weight they were very difficult to maneuver and it was necessary to lift the engine when turning into a street or around a building corner.[27]

It did not take long before American craftsmen began to build pumps similar to the Newsham and then to improve the equipment. In Philadelphia, Richard Mason constructed a double action pump operated by men on both ends, the hand brakes running longitudinally through the center housing. The discharge pipe was at the top.[28]

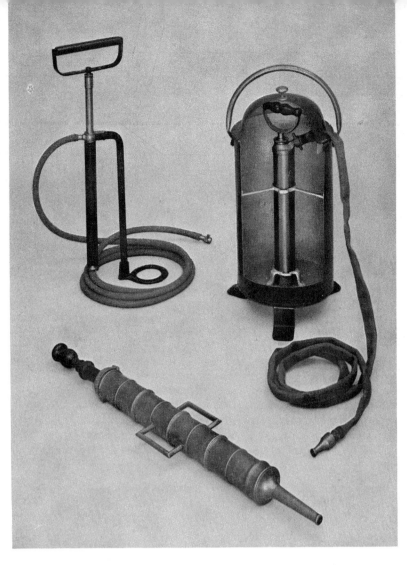

Three types of hand pumps used in Europe before colonies were established in America. The two stirrup pumps at top could only deliver a small amount of water. The squirt at bottom could be large enough to be held by two men and directed by a third, with water supply coming from a hand engine. (Fire Engines and Other Fire Fighting Appliances.)

In 1792 Richard Mason made this hand engine, said to be used by the Philadelphia Diligent Fire Company. (Strawbery Banke photo.)

In 1679, Boston purchased "ye engine lately come from England" and Thomas Atkins was appointed with a crew of twelve men to operate this engine. It was housed in a crude wooden shed which thereby became the first fire station in America and, since Atkins and the other men were paid salaries, Boston had the distinction of having the first paid fire department in the new colonies. By 1715 Boston had purchased six hand pumps of English manufacture. Three years later, insurance interests in the city formed mutual fire societies whose members were equipped with baskets and bed keys and, when they responded to fires, they tried to salvage goods and materials. These societies were the originators of some volunteer fire companies but actually they had no statutory authority or obligation. They functioned strictly as commercial firms protecting the interests of their sponsoring insurance companies.[29]

Wartime Use of Fire

Since the early days of mankind, fire has been used as a weapon of aggression, particularly in the total commitment efforts of war. In the history of America, military use of fire has been cruel, consistent and effective. In Colonial times this was particularly significant, because fire could rapidly devastate the combustible buildings used to house personnel and military supplies, and it was also effective in open land where troops would be foraging for game and vegetables.

In Massachusetts, in 1675–76, during the series of incidents known as King Philip's war, some 1,200 dwellings were burned and over 600 persons killed in the savage encounters between the colonists and the Indians.[30]

Other colonists received their baptism of fire during the French and Indian Wars (1754 to 1763) when England and France were trying to obtain sovereignty of Canada. When the occasion demanded, the Indians and troops used fire to wipe out or punish the settlements that did not assist them. The colonies in Virginia, the Carolinas and Georgia had similar conflicts with the Indians.

Just about the time when people in the colonies were dividing into Loyalist and Revolutionist groups, a situation occurred in Jamestown, Virginia, which was embarrassing to the Crown and also significant in aggravating differences between a royal governor and the colonists. In 1676, Governor Berkeley, appointed by the King, was the top official in Virginia. A young military leader named Nathaniel Bacon had been very successful in com-

Just before the Boston Massacre the town fire bell was sounded to bring citizens into the street. When the troops fired three colonists were killed instantly and seven others were wounded. Two of these died later. (Library of Congress Photo.)

bat with local Indian tribes and he returned to Jamestown following a series of victories which won him popular support. Following a series of encounters and arguments with Governor Berkeley, Bacon and a group of his supporters burned the Jamestown colony to complete destruction in 1676. When the Loyalists forces under Berkeley rounded up the offenders for punishment, Nathaniel Bacon died suddenly of some internal ailment, but subsequent hanging and other cruel treatment of his captured supporters by the British troops served to intensify resentments among Virginia colonists. This was to become more significant a hundred years later when the Revolutionary War was underway.[31]

Like most occupying troops, British military and naval personnel were ruthless in handling skirmishes, riots and military encounters with the colonists. The famous confrontation of the "Boston Massacre" in 1770 resulted following a ringing of the fire bell that brought a crowd of young, aggressive male colonists to hoot, jeer and ridicule British troops.

In the days preceding that tragic incident there had been numerous confrontations and challenges between the British soldiers and the ropemakers, the merchant sailors, and hundreds of belligerent young men and boys who moved in groups near the docks and the central part of Boston. It was on the night of March 5, at the end of a day when

15

The British ships Phoenix and Rose are attacked by American fireships on the Hudson River, New York, August 17, 1776. Two years later the American Lieutenant Colonel Silas Talbot commanded a fireship to attack the British ship Asia on the Hudson. Largest British ship destroyed by fire was the 64-gun HMS Augusta which caught fire and exploded in an attack on Fort Mifflin on the Delaware River October 23, 1777. Its following sloop of war, the Merlin, ran aground, was fired by its crew, and abandoned. (Official U.S. Navy photograph, courtesy of the Mariner's Museum, Newport News, Virginia.)

soldiers' drawn swords and cutlasses were matched many times against sticks and stones, with both sides suffering cuts and bruises, that the final confrontation boiled to reality.

One witness was quoted as saying that nine or ten soldiers came out of their barracks, each with a cutlass in one hand and a stick or bludgeon in the other. They struck several persons and headed toward an alley, despite efforts of their officers to get them back to the barracks. It was then that one of the Boston youths proposed ringing of the fire bell which was done promptly. Any alarm of fire in those days brought quick and concerned response from the citizens, and this alarm was no exception. People ran into the streets just in time for the confrontation between thirty or forty young people and a squad of British soldiers, led by their officer-of-the-day, Captain Thomas Preston. The rest is history — five Boston citizens were shot dead and seven were injured, at a time when emotions in the colonies were intensifying.[32]

In 1775, British ships sailed into Falmouth, Maine, to burn that community to the ground as punishment for aggressive actions of the colonists. Under command of a Lieutenant Mowart in the fleet of Admiral Graves, the crew bombed and burned 139 houses, 278 stores and other buildings, but the people of Falmouth stayed in their town and eventually repulsed the enemy.[33]

In that same year the people of Providence, Rhode Island, used fire for their own version of a Revolutionary Tea Party, as described in an old history of the fire department:

At noon on March 21, 1775, the town crier gave the following notice through the town: "At five o'clock this afternoon a quantity of India tea will be burnt in the market place. All true friends of their country, lovers of freedom and haters of shackles and hand-cuffs, are hereby invited to testify their good dispositions by bringing and casting into the fire a needless herb, which for a long time hath been highly detrimental to our liberty, interest and health."

About five in the afternoon a great number of inhabitants assembled at the place, where there were brought in about three hundred pounds weight of tea, by the firm contenders for the true interests of America. A large fire was kindled and the tea cast into it. A tar barrel, Lord North's speech, Rivington's and Mills and Hicks' newspapers, and divers other ingredients entered into the composition. There appeared great cheerfulness in committing to destruction so pernicious an article. Many worthy women, from a conviction of the evil tendency of continuing the habit of tea drinking, made free-will offerings of their respective stocks of the hurtful trash. On this occasion the bells were tolled, and the bright light of the fire illuminated the heavens for miles around, while the dense and suffocating smoke occasioned by the burning herb caused the spectators to withdraw from the immediate vicinity.[34]

Ship fires at sea and in harbor were very frequent during the Revolutionary War and sometimes became highly significant for the morale of either side, depending on the incident. At the start of the war the British fleet dominated many of the sea lanes but gradually American privateers and the American Continental Navy began to inflict severe damage on enemy shipping. The privateers were principally interested in capturing merchant vessels as "prizes" because some of these ships and cargoes rewarded the captors with large financial compensation when they arrived in port. This was very risky work on the high seas, and when the privateers or Continental Navy ships encountered British frigates and other armed vessels, the duels could be devastating. Fire and explosion fre-

In 1772 citizens of Providence, Rhode Island attacked the British Revenue schooner Gaspée in Narragansett Bay. After wounding the commander and dispersing the crew, the attackers looted the ship, then set it on fire. (Library of Congress Photo.)

quently occurred in these encounters. When such a sea battle was finished it was common practice to burn the losing vessel to complete destruction if it was not taken in tow as a prize.

A different kind of attack occurred off the coast of Rhode Island in March 1772. The British revenue schooner *Gaspée* arrived in Narragansett Bay and began aggressive patrolling to capture vessels that appeared to be smuggling cargo. It was reported that the British skipper, William Dudington, expressed intention to burn Newport, Rhode Island, "around the heads of its inhabitants."

On June 9 the packet boat *Hanna*, sailing from Newport to Providence, was followed by the *Gaspee* whose commander ordered the packet to stop for examination. The skipper of the packet refused and, with the help of a strong breeze, turned and sailed into the shallow water of the Bay. The *Gaspee*, following closely, suddenly ran aground and was stuck there until the next full tide.

Ashore in Providence, the word spread quickly that the hated vessel was trapped and before long a group of citizens organized to make a night attack.

Using eight long boats with muffled oars and crews armed with staves, stones and firearms, they rode down the bay for seven miles to where the *Gaspee* was stranded. In aggressive attack, they wounded the ship's commander, overpowered the crew, stripped the schooner of all valuables, and set the *Gaspee* on fire. It blew up when flames reached the powder magazine.[35]

Fire also became important in a famous ship battle off the coast of Britain in 1779. For more than a year, Captain John Paul Jones had been sailing in the shipping lanes near Britain capturing merchant vessels as prizes. On the morning of September 22 he was sailing a new ship, the *Bonhomme Richard* (named after Ben Franklin's *Poor Richard's Almanac*). Off the coast of Flamborough Head he encountered the new British frigate *Serapis* and thus began one of the most renowned battles in naval history. One of the first broadsides from the *Serapis* caused a violent explosion in the gunroom of the *Richard*, blasting two eighteen pound guns, killing most of their crews, shattering the battery and blowing a gaping hole in the deck. Very quickly, the British vessel had demonstrated its superior fire power and the commander called to Captain Jones to ask if he struck colors. Back came the famous answer "I have not yet begun to fight!" The battle continued for three and a half hours with both ships breaking out in fire frequently, sometimes so severely that it was necessary to stop the combat to subdue the fires, which several times threatened the explosives magazines of both vessels. Finally, after many casualties on both ships, the British commander struck his colors, not realizing how badly the American ship had been damaged. Captain John Paul Jones had a new prize vessel, but his own flagship, the *Bonhomme Richard*, was already sinking. There was hardly time to move the crew and the wounded onto the *Serapis* before the 9-ton ship *Richard* sank into the North Sea. But John Paul Jones had survived and he continued his famous naval career.[36]

When John Paul Jones brought his ship Bonhomme Richard into conflict with the British frigate Serapis, he found his ship being attacked by another American privateer, the Alliance. The reason for the attack was never determined but it was assumed to be a personal decision by the Alliance captain. (Smithsonian Institute.)

On June 17, 1775 when the British started the attack on Bunker Hill they set Charlestown ablaze, burning more than 400 buildings. This was a diversionary tactic as well as punishment for the local citizens. (Library of Congress Photo.)

In 1775, before the British marched on Bunker Hill in Boston, they set fire to some 400 dwellings across the bay in Charlestown. This was done more as a diversionary tactic than as punishment, but it served both purposes. The general intent of such burnings in this war, and later in the Civil War, was to break down the will of people residing in the area, and to create concern among the thousands of volunteer soldiers who were only serving for a few months before they returned home to their loved ones.

The use of destructive fire was a deliberate weapon to play on the emotions of these soldiers and cause them to desert or give up their military service.

As a countermeasure, fire and other stubborn forms of resistance were used to annoy, bother, and commit the occupying British troops to unplanned actions. Consider some of the suspicious events, mentioned in this report of the 1776 conflagration in New York City, at a time when the young, untrained American troops under command of General Washington were getting ready to meet the British professional soldiers in combat.

From an historical viewpoint there has never been agreement on whether this fire was started deliberately by the local residents, or whether it was of accidental cause, but subsequent actions of the British commanders indicated that New Yorkers were not exactly helpful for the occasion.

Under the command of General William Howe, the British troops were preparing to start a devastating attack on Washington's forces, which were in considerable disarray, with low morale after a series of recent defeats. Howe was planning to use New York as winter quarters while he began what was intended as a final mop-up of the colonial forces.

At that time, New York was a city of 20,000 population, primarily on the tip of Manhattan Island, but beginning to spread to the section later known as the East Side. Along the East River were the shipyards, and growth of the city was already planned for this area. The best residential center was down near the Battery, while the business district was around Hanover Square. Sidewalks were completed up to St. Paul's Chapel, but beyond that Broadway was just a dusty road.

Residents of New York infuriated the British on the night of September 21, 1776 by allegedly starting and aiding a conflagration that destroyed more than 400 dwellings. The British, about to attack the American troops commanded by George Washington, had planned to use the city for winter quarters. (Library of Congress Photo.)

About two o'clock in the morning of September 21, a fight started in a tavern and somehow a fire began. Very quickly sparks and flames were pushed along by gale winds from the southwest. Alarm bells were sounded from Trinity Church and Saint Paul's. Residents of the neighborhood responded and British sentinels called their superiors.

Then began a charade which sent the occupying troops into a fury. Manual fire pumps were dragged out and volunteers started them going as bucket lines formed. But suddenly one pump failed, then another, then a third. Half of the pumps could not provide enough pressure to make one good stream. Then someone began to screech that the buckets were no good — hundreds of buckets were found to have holes sliced in the bottom and water leaked out almost as fast as the buckets could be filled. With many of the regular New York volunteers across the river serving with General Washington, the fire fighting for this calamity became amazingly inept. The British officers, angry at being aroused at such an early hour, cursed the laboring civilians and sent their orderlies in flight to locate blacksmiths, shoemakers and machinists to make emergency repairs.

Meanwhile, the fire was racing into a major conflagration. Every building was involved between Broadway and Broad Street along the waterfront. Entire blocks were burning and the fire was already leaping across streets and igniting more structures. British officers tried to order the New Yorkers to fire fighting tasks but they were ignored.

When the soldiers tried to man the engines they were unsuccessful, and soon the whole fire fighting operation was a scene of bickering, angry shouting and confusion. The British ordered sailors and troop regiments to come ashore from warships in the East River, and for a while this manpower helped, but suddenly the fire leaped across Broadway, and headed northward. Instantly, more buildings were aflame.

When morning arrived, the conflagration was sweeping along both sides of Broadway unimpeded, until it reached the Hudson River, and by now fires were burning all over Manhattan Island. Trinity Church was totally demolished and for a few minutes it seemed that King's College, (now Columbia University), would disappear also, but students and professors were able to douse the small fires that started on campus buildings. Everything between Broad Street and Broadway down to the East River was completely destroyed.

Once more the conflagration burned through the night, finally ending along Fulton Street. It destroyed nearly 500 buildings, left thousands of persons homeless, and eliminated nearly ten million dollars worth of property.[37]

The British took savage revenge on any citizens who seemed to take active roles in starting new fires. Execution was swift, by hanging or bayonet stab. Major General James Robertson, military governor of the city, ordered his troops to arrest any likely incendiarist and very quickly 200 men were arrested on suspicion.

A view of New York from the harbor in 1776.

On September 23, General Howe wrote a letter describing the fire as "A most horrid attempt made by a number of wretches to burn the town of New York, in which they succeeded too well, having set it on fire in several places with matches and combustibles that had been prepared with great art and ingenuity."

General George Washington expressed a different opinion: "Providence, or some good honest fellow, has done more for us than we were disposed to do for ourselves," he wrote.[38]

One source quoted a New York merchant of the time, a David Grim, who saw the conflagration and left a record of the event. He said the fire started "in a low groggery and brothel, a wooden building," and aided by a southwest wind, swept up both sides of Broadway. He reported that the exact total of buildings destroyed was 493, of some 4,000 houses in the city.

Another resident, Captain Joseph Henry, was on a ship in the bay when the fire started and he witnessed the startling destruction. "When the fire reached the spire of a large steeple, south of the tavern," he wrote, "the effect upon the eye was astonishingly grand. If we could have divested ourselves of the knowledge that it was the property of our fellow-citizens which was consuming, the view might have been deemed sublime, if not pleasing."

And then, he reported, ". . . we clearly discerned that the burning of New York was the act of some madcap Americans. The sailors told us, in their blunt manner, that they had seen one American hanging by the heels, dead, having a bayonet wound through his breast. They named him by his Christian and surname, which they saw imprinted on his arm; they averred he was caught in the act of firing the houses. They told us also, that they had seen one person, who was taken in the act, tossed into the fire, and that several who were stealing, and suspected as incendiaries, were bayonetted."

And he concluded, "The testimony we received from the sailors, my own view of the distant beginnings of the fire in various spots, and the manner of its spreading, impressed my mind with the belief that the burning of the city was the doing of the most low and vile persons, for the purpose not only of thieving but of devastation."[39]

One of the major changes in military conduct of the Revolutionary War occurred when Lord Charles, Earl of Cornwallis, commanding the British troops in the Carolinas and Georgia, was aggravated because of guerrilla tactics assumed by the American militia. In 1780 he had occupied Charlotte, North Carolina, after defeating all attempts of the American troops to confine and defeat his forces. To Cornwallis and his supporting officers, it appeared that the war could be ended quickly, especially after the defeat of troops under the American General Horatio Gates. However, following this encounter, General George Washington replaced Gates with Major General Nathaniel Greene, a brilliant tactician. Greene had several outstanding officers in his command: Brigadier General Francis Marion, later called the "Swamp Fox"; Brigadier General Daniel Morgan; and two local leaders, Thomas Sumter and Andrew Pickens. With these officers commanding local guerilla troops, Greene began to disrupt the formalized military maneuvers of the British troops. He succeeded so well that Cornwallis, in irritation after the battle of Cowpens, South Carolina, burned "superfluous baggage and all my waggons, except those loaded with hospital stores, salt and ammunition. . . ." His purpose was to strip his infantry troops to the least baggage so they could pursue American troops "in a fast march."

When General Greene heard of this deliberate burning of supplies he said, "Now he is mine!", and began a classic maneuver of guerrilla warfare that has been repeated in modern military conflicts. Greene gradually withdrew his forces under the steady pressure of the British troops, but only allowed his rear echelon to partially engage the enemy, then to break contact, as the main American forces withdrew in controlled manner through South Carolina, North Carolina and into Virginia. Meantime, local partisans and Greene's cavalry troops were burning and destroying the supplies and land area *behind* the British troops that might be used for food supply, in effect isolating the enemy from forage or other essentials.

Greene's army finally made a stand at the Dan River, between North Carolina and Virginia in February 1781, at a time when the forces under Cornwallis were exhausted and depleted by at least five hundred of their troops after the tantalizing engagements with the guerrillas. The American cavalry, under command of General Henry "Light Horse Harry" Lee (father of the later Confederate General Robert E. Lee), made devastating attacks on the British cavalry and infantry troops.[40]

One of the subordinate officers to Cornwallis was Lieutenant Colonel Banastre Tarleton, who, in the previous year, had played a decisive role in the burning of Bedford, New York, and later in the savage battle of Waxhaws, on the border of South Carolina. It was then that the term "Tarleton's Quarter" was created, because Tarleton ordered his men to kill all surrendered opponents without mercy, an action considered a violation of military protocol of that time. It was not until the surrender of Cornwallis at Yorktown, Virginia, that Tarleton was finally captured by American troops, but his legion had been very active under command of Cornwallis. In January 1781, Cornwallis ordered him to push the American troops commanded by the American Brigadier General Daniel Morgan to the utmost. Morgan, following the tactics of Marion and Greene, retreated gradually to Hannah's Cowpens, near the North Carolina border. It was here that Morgan slaughtered Tarleton's troops in one of the most decisive battles of the war. [41]

A fire in New York in the 1770s. Note men rocking brakes of
the hand engine at left, the well supply at right, and the rather
dispirited bucket passers.

Another Colonial times fire is satirized in this sketch. Second
fire company is arriving on street at right while calamities
occur in the foreground.

Fire Marks —
and a Long Search[1]

The Insurance Company of North America has an
outstanding collection of fire marks dating from 1752.
(See Appendix for address — Ed.)

The Great Fire of London taught its lesson. During the very next year, 1667, the first fire insurance office was established, by Dr. Nicholas Barbon. Shortly after its reorganization in 1680, the Fire Office, as it was known, formed a "company of men versed and experienced in extinguishing and preventing of fire." This was the first fire brigade in London and — if we except the guards at Kingsay, now Hangchau, China, to which Marco Polo referred — the first organized body of firemen that we know of anywhere since the time of the Roman "Matricularii" and "Cohortes Vigilum."

Naturally, the Fire Office created its brigade for just one purpose — to protect the property of those insured by the Office, thus preventing losses which the Office would be called upon to pay.

In order that the firemen might readily distinguish insured from uninsured properties, the Fire Office adopted as an identifying mark a metal plate in the form of a Phoenix rising from the flames. This Fire Mark was nailed up in a prominent spot, about level with the second story windows, on all houses insured by the Office.

The idea of fire insurance took rapid root; and, with the advent of rival companies, the original Fire Office soon found it necessary to adopt a more distinctive name. Accordingly, in 1705, it assumed the name of the Phoenix Fire Office, a name obviously suggested by its Fire Mark. Under this name, it continued in existence until some time in the early part of the eighteenth century.

Each of the other fire insurance companies had, of course, its own fire brigade and its own Fire Mark. As a result, when an alarm was raised, all the brigades responded on the double-quick. On arriving at the scene of the fire, they looked for the Fire Mark. Whereupon all except the one brigade whose emblem appeared on the house either turned tail and went back to bed or, more frequently, remained nonchalantly in the background to cheer and jeer the firemen of the rival "office."

If the building had no insurance — and no Fire Mark — the hapless owner had to depend on the ineffectual buckets of his neighbors and friends. Needless to say, he usually got the insurance — and the Fire Mark — as soon after as possible.

All of the early Fire Marks in London were of lead with the number of the insurance policy usually stamped on. They were brilliantly painted, generally in red and gold; and, as the amount covered by a policy was usually very small, frequently the marks of five or six different companies were found upon the same risk, which must have given the old-fashioned houses a rather gay appearance. A rhyme published in 1816, referring to a certain overdressed English lord, aptly says:

"Fir not e'en the Regent himself has endured
(Though I've seen him with badges and orders all shine
Till he looked like a house that was overinsured) —"

Gradually, as business increased and many buildings became insured in more than one company, the brigades began to unite and attend all fires. From then on Fire Marks lost their original significance, and were finally supplanted by what collectors now call Fire Plates, usually made of copper, tin or iron, and used solely as advertisements. So strenuous was competition among the various companies, and so eager were they to have their Plates displayed, that one large company used to keep a man with a ladder constantly employed in putting up its Plates wherever space was available, regardless of whether it covered the risk.

Finally, in 1833, the fire brigades were all amalgamated into one company and in 1866 were turned over to the City of London. This was the beginning of the "Metropolitan Fire Brigade of London," and the end of the picturesque Fire Mark in that city.

The early American colonists, with their pioneer aggressiveness devoted themselves to the active phases of fire combat, neglecting fire insurance in spite of the fact that insurance was well established by that time in Europe. Bucket brigades were in operation as early as 1696; a fire engine was brought over from London in 1718; two more were imported in 1731; and, finally, the following year, a Philadelphia mechanic, Anthony Nicholls, built a very successful model which "played water higher than the highest in this city had from London."

In 1736, Benjamin Franklin and several other prominent Philadelphians established the first organized fire brigade, which continued as the Union Fire Company for over eighty years. Other companies quickly followed until, in 1752, there were no less than six in Philadelphia, with an aggregate membership of 225, employing 8 engines, 1,055 buckets and 36 ladders.

Each of these old fire companies had its own name, insignia and full regalia. A fireman was a sight wondrous to behold — resplendent in high hat, gaudy cape, huge belt and buckle with his insignia engraved thereon. Fully equipped he proudly bore horn and axe, leathern bucket and linen bag (to protect salvage from looters).

Originally, these fire companies were organized simply for mutual assistance. Each member paid for his own buckets and other paraphernalia. Their efforts were mostly confined to protection of their own members' homes. Many prominent citizens belonged. Later, volunteer companies were formed on a purely freelance basis, putting out whatever fires they were able and depending for pay on the bounty of the owners.

In 1752, the first American fire insurance company was formed, again under the inspiration of Benjamin Franklin. This was the Philadelphia Contributionship for the Insurance of Houses from Loss by Fire. Only five weeks after organizing, they ordered a hundred Fire Marks from John Stow who was, a year later, to recast the historic Liberty Bell. The Mark consisted of four leaden hands, clasped and crossed a la "My Lady to London," and mounted on a wooden shield. From this mark, the company became popularly known as the "Hand-in-Hand." In 1792, the Insurance Company of North America was formed — the first American fire and marine insurance company. Other insurance companies sprang up with new Marks.

Contrary to the English custom, these companies did not maintain fire brigades of their own but relied on the volunteer fire companies already flourishing. The American Fire Mark had, therefore, a very different mission. In the first place, an insurance company's Fire Mark on a house was intended to discourage malicious arson by showing that the owner himself would not greatly suffer if the building were destroyed. Secondly, the Fire Mark stood as a guarantee to all fire brigades that the insurance company which insured the house in question would reward handsomely the brigade extinguishing a blaze on the premises. Stimulated by higher and more certain compensation, competition among the fire brigades waxed hot, soon reaching a stage of bloody noses and blackened eyes.

In 1871, Philadelphia inaugurated its paid fire department. From that date, Fire Marks lost their usefulness and were discontinued.

The story of how just one prized trophy was acquired is a real romance.

In 1914, an authority wrote, concerning the Insurance Company of North America, "this company had two Fire Marks. The first was adopted December 8, 1794, as 'a wavy star of six points, cast in lead and mounted on a wooden shield.' I believe no specimens are in existence."

Later the company adopted the "Eagle" Fire Mark of lead on wood of which not more than two specimens have survived the ravages of time. A well-preserved specimen is in the collection.

In 1915, another wrote, "The only specimens of the Fire Marks of this company now existing are those of the 'Eagle' variety."

In 1928, another wrote, "No specimens of this Mark (the 'Star') are known to exist, the last one located having appeared on the building at 229 South Front Street, Philadelphia, in 1879."

The hunt for the "Star" began when a noted collector and authority discovered a tell-tale stain on a brick house in Race Street, near Second, Philadelphia. This stain was exactly the same shape as the shield of the lost "Star" Mark, as shown on an old etching of the company's first office building where the artist had outlined the "Star" Mark over the doorway. Tracing back from the owner of the building to his father and grandfather, the wooden shield was at last located in the possession of a carpenter. What a disappointment, though, to find a shield of the right shape and age — but nailed on it a not at all rare "Fire Plug" Mark of the Fire Association of Philadelphia! Closer examination showed under this more recent Mark a faint outline which proved that the shield originally carried the "Star"!

With whetted appetites, the searchers sought for clues. The best was a vague recollection by the grandfather of having sold the metal "Star" to a "secondhand dealer from Baltimore." Then to Baltimore and a systematic combing of every antique and secondhand shop in the city. At length, on the outskirts, they found a dealer who said, "Sure, there's a star here. Been here for years — long before I bought the shop."

Chemical tests proved the age of the lead; and, what is more, the "Star" fitted the old wooden shield precisely — nail hole for nail hole, line for line! Today this one North America "Star" stands unique, the rarest of all Fire Marks.

Burning of merchants' exchange building in New York City, part of devastating conflagration of 1835. (Smithsonian Institute.)

1801-1860

Chapter II

". . . Immediately on notice that a fire hath broke out, assemble on the spot and render the most effective service. . . . Each and every householder, occupier of a store, or shop within limits of the town . . . furnish himself or herself with two leather buckets to be made use of in case of fire . . ."[1]

The 19th century began as an exciting era of change and expansion for the newly formed states and Federal government. Thousands of settlers were arriving from Europe; many new businesses and industries were being started; the American merchant fleet and naval vessels were gaining respect and recognition from the older countries; and all along the eastern coast of the new country cities and harbors were developing. But the century brought a full share of disaster to balance the excitement of discovery, innovations, and rewarding enterprise.

Expansion of shipping and trade led to conflict with the British and the consequent War of 1812. The practical application of steam power in industry and for fire department pumpers, and the use of horses, came at a time when the states were beginning the savage sectional conflict that was to evolve into the Civil War. The rapid expansion of cities, with most structures built of wood framing with wood-shingle roofs, created ideal conditions for the conflagrations that would shock the nation in the latter part of the century.

Fire companies were becoming better organized and the innovation of steam power soon would replace the tedious strain of keeping pumps in action by manual labor. Horses and self-propelled apparatus would not be accepted by the volunteers until about mid-century, and then the first of the fully paid fire departments would bring even greater change in municipal fire protection.

But many fires had to occur with consequent loss of life, tragic suffering, and huge destruction of dwellings and business properties, before the era of modern fire protection was to begin.

Again, arson, or incendiary burning, was to loom as a major fire problem, particularly as disputes developed between states prior to the Civil War. Hundreds of barn, dwelling and bridge burnings occurred along the borders of Virginia, Tennessee, Kentucky, Missouri, Kansas and Nebraska as local and sectional rivalries developed between prior settlers and new settlers.

A famous character at this time was James H. Lane who had been elected as a U.S. Senator from Kansas. In addition to his senatorial role, Lane had the commission of brigadier general and led some of the Federal "Jayhawker" troops from Kansas who swept along the Nebraska border in vicious raids. When troops under his command laid waste to Osceola, Missouri, one report described Lane as "looking like some Joe Bagstock Nero fiddling and laughing over the burning of some Missouri Rome."

But the moods and actions of the people involved were no laughing matter to Thomas Jefferson when the hot disputes of the Missouri Compromise arose during his Presidential term. His agony of decision was best summarized by his famous "firebell in the night" statement. (See page 37.)

Until 1800 no fires of consequence had occurred in Washington, designated to be the nation's capitol, but in that year a very severe fire destroyed a three-story building occupied by the War Department. Despite the sounding of an alarm and quick work of local citizens, the building and contents were ruined — a loss severely felt by the national government because many valuable papers, impossible to duplicate, were destroyed. Only one volume was rescued; in it were recorded contracts and deeds of land sold to the United States, together with papers of the accounting officer.

Disastrous Conflagration —

Just before 5 o'clock last evening, says the New York Dispatch, a fire was discovered in the National Theatre, corner of Church and Leonard streets, and in a very few minutes the whole building was enveloped in flames. So rapid was the progress of the fire that it was found impossible to save but a small part of the properties of the house, and therefore the loss to both the manager and actors, must be very great.

Soon after the fire had burst through the roof of the theatre, it communicated to the African Zion Methodist Church, on the opposite corner, and also to the dome of the magnificent French Catholic Church, situated on the corner of Church and Franklin streets, adjoining the Theatre. The African Church, having a wooden roof was soon on fire in every part, but the thick tin roof of the French Church withstood the flames for a considerable time, and it was a strange and imposing sight to see the dome in flames while the church itself remained uninjured. The wood work of this splendid pile at length ignited from the immense heat from the theatre, and the fire spread rapidly throughout the interior, destroying the elegant pictures, gildings, trappings of the pews, &c. &c., as scarcely any thing from within had been rescued.

About this time the rear wall of the theatre fell outwards, from which catastrophe, another church was set on fire, viz. the New West Dutch Reformed Church, under the pastoral charge of the Rev. Mr. Hunt. This Church was situated in Franklin street, about 150 feet below the French Church, and the rear of it was adjacent to the theatre. As soon as the flames had communicated to it, they spread much more rapidly than in the French Church, as in the course of a very few minutes it was in flames throughout. Besides the entire destruction of these three churches, and of the theatre, there was no considerable damage done.

A two story wooden house adjoining the theatre on Leonard street, occupied by Julia Brown, a notorious prostitute, was partly destroyed, the roof and attic story being burnt.

One hundred and eight persons attached to the Theatre, are deprived of employment by the sad catastrophe, and many of the actors have suffered severely in the loss of their wardrobe.

The fire originated in the gas-room, under the stage, near the pit lobby. A new gasman had just been employed, and, from some mismanagement, had left the gas running in such a manner as to fill the room, when it ignited suddenly from a lamp which was kept burning, and the flame shot up and communicated with the scenery almost as suddenly as though the house had been struck by lightning.

It is unnecessary to say that the rumors that it was the work of an incendiary are unfounded.

Baltimore Clipper, September 26, 1839.

Seven months later, on July 20, 1801, a fire in the building of the Treasury Department burned out several rooms, destroying ledgers of the Quartermaster and Commissary Department, accounts for the Bank of the United States covering sale of stock claims during the War of Revolution, and for public supplies from 1797 to 1798.[2]

In 1843 an explosion in the Washington Navy Yard involved several buildings and threatened the powder house, but quick actions by two members of the Anacostia Volunteer Fire Department prevented another major explosion.

On March 15, 1845, fire involved a room of the National Theater and spread quickly through the entire structure. The night was extremely cold and fire fighters had a great deal of trouble with frozen hose and hydrants. For a while it appeared that the fire was going to spread into a major conflagration, but after destroying the entire block, flames subsided and it was controlled.

In 1848, a destructive fire occurred in a building across from the White House. This contained a number of important papers which were destroyed despite valiant efforts of local fire fighters to save them.

On December 24, 1851, a major fire started in a small hotel within the city. Again, extremely cold temperatures became significant, as fire fighters tried to handle frozen hose and apparatus. They had finally controlled this blaze, when they received word that another fire had quickly involved the Library of Congress, specifically the library of former President Jefferson, which then was housed in the central wing of the Capitol building. It gained great headway before fire companies responded and fire fighters were unable to save anything except four or five president's portraits and the original Declaration of Independence. Rare marble sculptures, a number of important medals, and many magnificent works of art were destroyed. The entire Library of Congress contained some 60,000 volumes most of which were destroyed. The Anacostia Fire Company attempted to get a suction pumper into the Rotunda but it was too large to pass through the door. A smaller hand engine from the Columbia Fire Company was able to enter and the consequent interior fire fighting proved effective.[3]

(The nation's Capitol had other fires in the following century. On November 6, 1898 a gas explosion and fire severely damaged the Supreme Court chamber, the room formerly used by the U.S. Senate.

On July 2, 1915 the Senate reception room was damaged by the explosion of a homemade bomb, placed there by a man who was upset by private sales of American munitions to the Allies during World War I. And on March 1, 1971 a bomb explosion caused considerable damage on the ground floor of the Senate wing. — Ed.)

The Crystal Palace fire in New York, October 5, 1858. There were about 4,000 exhibitors and about 10,000 people were expected to attend an evening concert in the building. Remarkably, no one was killed in the fire which later was determined to be incendiary.

Another notable fire occurred in New York City in 1835. This began on the cold evening of December 16 and, before it was extinguished, it consumed nearly 700 mercantile buildings, with the contents valued somewhere between twenty and forty million dollars. The fire started at 9 p.m., swept through the close, combustible buildings with their extreme fire load of mercantile products, and defied most control efforts of the New York City Fire Department. Mutual aid responded from volunteer engine companies in Philadelphia and New Jersey but, because of freezeup of hydrants, pumps and fire hose, and other complications, control efforts were relatively ineffective. Kegs of gunpowder were used to destroy buildings, thus creating a fire break to interrupt progress of flames from one combustible section to the next. Hogsheads of vinegar were used effectively to protect at least one building exposed to the sweeping flames. By December 19, the fire diminished and was near to extinguishment, but it had destroyed buildings within a quarter-mile square in a very valuable district of the city. This catastrophe prompted the mayor to appoint private patrols to cover the city in event of a secondary fire outbreak. Several companies of military personnel and marines were appointed to protect property within the city.[4]

In August 1858, the Crystal Palace was destroyed. This was a world-famous building constructed to house the World's Fair exhibition which opened in 1853. The Crystal Palace covered an entire block on Forty-second Street. It was constructed of some 1,250 tons of iron and 39,000 square feet of glass. Subsequent to the World's Fair, and at the time of the fire, it contained many priceless works of art and was described as "the most tasteful ornament that ever graced the metropolis." But when fire started within the structure, it spread so rapidly that dome and roof collapsed within twelve minutes and the contents were completely destroyed. Approximately 2,000 spectators were reviewing the exhibits when the alarm was sounded; miraculously, they were able to escape to safety through doors on the Sixth Avenue side of the building. Even though the fire department responded promptly to the alarm, the intense involvement of the structure made fire control efforts hopeless. This fire received a great deal of notoriety because of the attractiveness of the building and the irreplaceable loss of its artistic contents.

Two years later, in 1860, fire destroyed a tenement building on Elm Street in New York killing twenty impoverished people who died in the flames

or leaped to destruction. This incident caused the Common Council to pass a strict fire protection and building ordinance requiring fireproof stairs or fire escapes, as follows:

In all dwelling houses which are built for the residence of more than eight families, there shall be a fireproof stairs, in a brick or stone or fireproof building attached to the exterior walls, and all the rooms on every story must communicate by doors; or if the fireproof stairs are not built as above, then there must be fireproof balconies on each story on the outside of the building, connected by fireproof stairs, and all the rooms on every story must communicate by doors. If the buildings are not built with either the stairs or balconies as above specified, then they must be fireproof throughout. All ladders or stairs from upper stories to scuttle or roofs of any building shall, if moveable, be of iron and if not moveable may be of wood; and all scuttles shall be not less than three feet by two feet.[5]

A Nathaniel Currier lithograph of the 1845 fire in New York as seen from Bowling Green.

The "Great Fire of 1835" in New York City destroyed nearly the entire business portion of the city — about 13 acres, 700 buildings. Loss was estimated at 20 to 40 million dollars, a mammoth sum in those days.

Broad Street explosion during the 1845 fire.

About 600 employees were in the publishing buildings of Harper & Brothers when fire erupted December 10, 1853. The fire started when a plumber dropped a match into a pan of flammable liquid. Everyone escaped from the building but loss was in the vicinity of a million and a quarter dollars, a huge financial setback for the publishing company.

About twenty volunteer fire fighters were buried in the collapse of the Jennings building April 25, 1854. Note the number of men operating the brakes of the hand engine in the foreground and the crowds near the hose lines.

Disastrous Fire in New York
— Fall of a Wall —
Loss of Life

On Tuesday evening last a fire broke out in the building situated at No. 231 Broadway, New York city, occupied as the clothing and furnishing store of W. T. Jennings & Co., a few doors above the Astor House. So rapidly did the fire spread that, although the alarm was promptly responded to by the Fire Department, the flames rose high above the roof before any streams could be brought to play, casting a brilliant glare over the Park and lower part of the city. Some thousands of persons gathered in the vicinity; the Hose and Engine companies arrived in quick succession, and soon the neighborhood became crowded. Despite the exertions of the firemen, the flames spread to the adjoining building, No. 233, the lower part of which was occupied by W. A. Batchelder's hair dye and wig establishment, and the roof of building No. 229.

It was while the firemen were straining every effort to prevent the flames from spreading to the adjoining buildings, that an accident occurred attended with a loss of life and limb more disastrous than any we have had to record since 1845.

In obedience to an order to "light up the hose" of Engine Company No. 21, a number of firemen entered the cellar of building No. 231, and while there, a partition wall gave way, carrying with it a portion of the first story flooring, on which was a heavy safe, and burying several men beneath the ruins. A wild cry of horror arose from those outside, as they saw the dreadful fate of their comrades, and instantly every effort was made to rescue them from the ruins. The scene now became most harrowing and exciting.

The Firemen were distracted between efforts to rescue their comrades and subdue the flames; while to the crowd beyond the calamity appeared even greater from the confusion and uncertainty which surrounded it.

At length the fire was subdued, and undivided exertions were given to removing the ruins.

By 1 o'clock, a total of ten or a dozen persons were taken from the cellar, four of whom were dead, and several so badly injured as to be removed to the Hospital. In rushing in to save their comrades several others were injured. From the return given of the list of persons killed and wounded, the N. Y. Times makes up the following:

Dead	4
Missing	2
Fatally injured	2
Badly injured	14
Slightly injured	11
	—
Total	33

The N. Y. Courier and Enquirer thus states the cause of the fire:

It appears that the rear wall of the main building was supported by an iron arch with the extension of the house in the rear, and when the pillars which supported the arch were burnt away the whole wall fell in, crushing many who were then inside. When we visited the scene of the disaster the firemen were breaking holes through the basement walls of the adjoining buildings, and endeavoring to remove the ruins so as to save their gallant brethren who had so bravely perilled their lives.

The walls fell twice. The rear wall fell first, burying several, and while releasing those the side wall fell, burying the rescuers.

Great complaint is made in regard to the unsubstantiality of the building. It was found to be a mere shell, built upon the "cheap principle," and to this is owing the disaster which we now record.

The Daily Times says: The roof of the large building No. 229 was burned, and the whole interior flooded. The building is fully insured. The papers, &c., in the different offices, were completely saturated, and many of them are rendered worthless.

The flames also spread to building No. 235. The roof and the upper story were considerably damaged; and the interior of the building was also flooded with water.

A postscript in the Times says:

SIX MORE DEAD! — Just as we are going to press, 2½ a.m., a strong body of firemen are busily at work in removing the ruins. A son of Coroner O'Donnell, aged 21, and attached to No. 42 Engine, has been spoken to in the ruins. He is buried under the rubbish at a depth of 15 feet. He says that *six men* are lying near him *dead*! He is as comfortable as the circumstances would admit, but there is fire very near him, and the rescuing party are obliged to turn a stream of water toward him, from time to time.

Several of the firemen reported injured were hurt in endeavoring to rescue their comrades from an awful fate.

Alderman Howard, Assistant Engineer, was on the upper floor of the burning building when the partition wall fell with such terrible effect. He succeeded in springing from the second story window, apparently without receiving any injury.

Zophar Mills, ex-Assistant Engineer, had a narrow escape. He, also, was compelled to leap from the second story to the ground, but was not injured.

North American and United States Gazette,
April 27, 1854.

The bursting of a two hundred horsepower boiler demolished this six-story Hague Street building in New York, February 4, 1850, killing sixty-three of about a hundred persons who worked there.

San Francisco experienced a number of fires after 1848, when it became an established city. But in a very short time, when gold was discovered in California and nearby states, the influx of new residents and the rapid expansion of San Francisco and surrounding communities created entirely new municipal problems.[6]

In January 1849, one of the major hotels in the city burned to destruction, and six months later the ship *Philadelphia* burned at her moorings. A more serious fire, however, was in the last week of 1849.

It is a matter of history that San Francisco was a "wide open" community with saloons, gambling halls and quick-profit making establishments. This fire began in the early hours of morning the day before Christmas. Despite the efforts of many citizens, the fire spread very rapidly north and south, destroying all of the combustible buildings in its path. Fire fighting efforts generally proved useless, particularly because there were no organized efforts to establish fire control. Explosives were used to destroy some buildings and provide fire breaks.

Other buildings were pulled down for the same purpose, and the rest of the city was saved by such desperate efforts.

On May 4th, 1850, fire again swept through a number of combustible buildings, and subsequently reports indicated that this was another fire developed by arson. This particular blaze started at four o'clock in the morning in almost the same location as the 1849 fire. It spread very quickly, sweeping over two complete blocks.

On June 14, 1850, a fire started in a bakery behind the Merchants Hotel within a few hours destroyed all buildings within a large area leading down to the bay.

At that time, San Francisco consisted of many temporary structures, including tents and hastily built wooden buildings, which were extremely vulnerable to fire. Very quickly, several other arson fires of the same magnitude created an equally disastrous effect for the city, leading to formation of vigilante groups who patrolled the city watching for arsonists.

Here is a lithograph of the Great Conflagration in Philadelphia, Pennsylvania that began July 9, 1850. This illustrates an explosion of gunpowder in a Water Street building. More than 500 houses were destroyed by the fire, fifty-seven persons were killed; 115 were wounded; nine were drowned; and loss of property was estimated at more than one million dollars. (Library of Congress Photo.)

On the night of May 3, 1851, San Francisco suffered another disaster, also reported as a result of arson. It began at eleven o'clock at night and, within a few hours, involved the entire business district of the city in a massive fire, visible in Monterey, one hundred miles away. Uncounted lives were lost in this conflagration which destroyed between fifteen hundred and two thousand buildings. It wiped out eighteen entire blocks and caused damage exceeding twelve million dollars. Apparently, the only reason it stopped was because there was nothing left in its path to be burned.

This major disaster was followed by another fire on Sunday, June 22, but fortunately this was controlled fairly well, after it had destroyed buildings within fifteen blocks.

Saint Louis, Missouri, also was in the process of expansion, and, in 1849, it suffered an epidemic of cholera which arrived with boatloads of immigrants and gold-seekers heading for California. When the epidemic was at its height, the "Great Fire" of May 17 occurred along the Missouri River. Here the steamer *White Cloud* suddenly broke out in flames where she was moored with other riverboats. Very quickly the fire involved twenty-three steamboats, then a group of buildings along the waterfront. Before long, 400 buildings in fifteen city blocks were burning, with losses later estimated at more than five million dollars. "With the heaven in flames and the people in terror, the fire took hold on Front Street and swept westward," said one report.

A Currier and Ives rendition of the great fire in Saint Louis, Missouri, May 17, 1849. Twenty-three steamboats and cargoes were destroyed along the waterfront; hundreds of buildings and other land properties were wiped out in the conflagration. (Library of Congress Photo.)

Once again, fire officials decided to use explosives to destroy buildings in the path of the blaze, but a tragedy resulted. One of the department's officers, Captain Thomas B. Targee, who commanded the Missouri Company, had been fighting the fire for many hours. He had entered three of the six buildings that were destroyed by explosives, trying to determine the progress of the flames. He was last seen entering the fourth building, cradling a keg of gunpowder on his shoulder. A few moments later there was a violent explosion that leveled the building and killed Targee. The cause of the premature blast was never determined.[7]

It is not surprising that this became a nation of wooden houses. When brick was used for exterior walls the whole inside was of wooden construction. Not only did we use wood extensively, but we began to use it poorly, that is, from the fire protection point of view. Older structures were built of heavy timbers with mortised joints which was extremely wasteful of lumber but slow burning. It was, of course, a very slow method and it was not long before joisted construction was developed which made good use of wood from the point of view of structural strength and which made for much quicker and easier construction. But from the fire protection point of view that change was not to the good. Joisted construction burns so rapidly that it merits the description "quick-burning" given to it by fire underwriters and fire authorities. Brick buildings with the interior filled with wood-joisted construction were not much better than those of entirely wood or frame construction. The effect was something like a stove filled with kindling wood and the fact that the outside walls might not burn did not add much to the safety of those inside.[8]

Silver trumpet belonged to the chief of the Detroit, Michigan Fire Department. (Detroit Fire Department Photo by Captain J. A. Mancinelli.)

Silver trumpet of Chief Ashfield of the Toronto, Ontario Fire Department.

Other Fire Department Developments

The 19th century brought some of the most significant changes in fire department methods of fire control and general fire protection. The capital cities and principal communities in the new states were beginning to organize volunteer fire departments and to obtain the new pumps and other equipment that was becoming available.

In 1808, the firm of Sellers and Pennock in Philadelphia introduced leather fire hose seamed by copper wire rivets. This was the beginning of an entirely new concept of fire fighting in which water could be relayed for hundreds, perhaps thousands of feet, with sufficient pressure to attack a fire effectively from many positions. Subsequently, fire departments requested hose of oak-tanned leather with double riveted seams and splices made with thirteen rivets of Number 7 wire. When finished, the hose, with carrying loops and rings, weighed approximately eighty-four pounds for each fifty-foot length, exclusive of metal couplings. It was required to withstand a pressure of 200 pounds per square inch. It was not long before suction hose was developed, made of short metallic cylinders covered with canvas, leather, or spiral wire.[9]

In the same year the first working hydrant was installed in New York City, attached to an underground wooden water main. Earlier, four hand pumps similar to the Newsham had been built for Philadelphia and just about that time fire escapes were used for the first time in that city. Three years later, water was diverted from the Schuylkill River into "trunks" in the street to be available for fire fighting. The Philadelphia Hose Company was formed and quickly placed in service a hose carriage with the riveted leather hose of Sellers and Pennock. In 1806, a large fire occurred in the city destroying thirty-two buildings. More hose was purchased and inspection of hydrants was begun. By 1810 there were nine hose companies; the engine companies were also provided with hose to be attached to the increasing number of hydrants.

Volunteer fire companies protected the city of Philadelphia for 120 years, from 1750 until December 29, 1870, when the Select and Common Council of the City enacted a special ordinance to establish the paid fire department. Many of the volunteer firemen were selected to be members of this new department and some of the locations of the volunteer fire companies were used for the new stations. By that time, Philadelphia had a variety of apparatus, including Amoskeag and Neafie and Levy steamers, and hose carts built by Gardner and Fleming. The changeover to the paid department was not accomplished easily, since the volunteers had formed nearly 100 organizations scattered throughout the city and they wielded very strong social and political influence. However, the change was gradually accepted.[10]

Interesting changes were underway in other cities. In Charleston, South Carolina, the fire department received assistance from policemen who attended fires armed with wooden staves to keep order among the crowds. The police were also required to pass hand buckets down the line to supply engines and for many years the police department had its special fire squad.

Charleston maintained two explosives magazines, each with fifty charges of powder to be used for blowing up buildings in major fires. Prior to 1859, a powder cart was hauled around by hand to be used in this "saving" destruction. When charges of powder were about to be exploded a trumpet call would be sounded to warn people to stand back from the area.[11]

Among other interesting changes in the first part of this century were the following:

In 1819 a special fire patrol was organized in Philadelphia to perform salvage work at a fire. They used large baskets for this purpose.

In 1821 a patent was issued for rubber-lined, cotton-jacketed fire hose to replace the riveted leather hose used in Philadelphia.

In 1840 the independent, direct-acting, steam-power pump was invented by Henry R. Worthington, who subsequently formed the Worthington and Baker Pump and Machinery Company in Harrison, New Jersey.

". . . this momentous decision, like a firebell in the night, awakened and filled me with terror . . ." President Thomas Jefferson referring to the Missouri Compromise.[24]

In 1852 a patent was issued for the first sprinkler perforated pipe system designed by James Pichen Francis, proprietor of Locks and Canals at the Merrimac River in Lowell, Massachusetts. This probably was the first recognized installation of fixed fire protection equipment, but the full, expansive development of sprinkler systems did not really get underway for another twenty years.

From this time on, inventions and other fire protection developments occurred at an increasing pace, but there were some retarding influences that had to exist until other progress was a reality. Sprinkler systems, for example, had to be delayed until metal piping and substantial pumps were developed. The fusible sprinkler head was still a few years away; alarms, post indicator valves, siamese connections, and accurate hydraulic data would not be available until the twentieth century. The same slow progress affected every type of fire protection equipment.

Riveted leather hose of the type used in the 1850s. Note leather handle and curved grips for securing the non-threaded coupling. (Smithsonian Institution.)

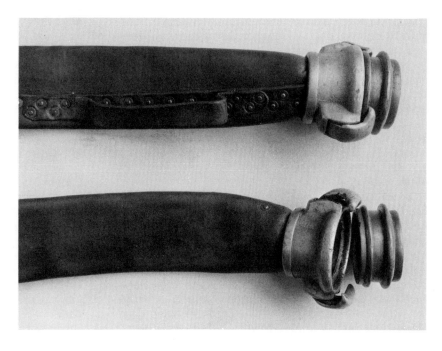

Rendition of burning of Washington, DC by British in War of 1812. At upper right are General Ross and the British army. Directly below is the American Flotilla in flames. City of Washington is in upper left. Dock Yard and rope walk just below. The Potomac River is in center where bridge is being destroyed. President's home, Senate building, and Treasury are in upper center, in flames. (Library of Congress Photo.)

The War of 1812

The first signs of a war that nobody wanted developed when American shipping continued in conflict with British interests, and the British countered by capturing and impressing crews of these ships. On land, there still existed strenuous resentment toward Britain, dating from the Revolutionary War.

The emotional issues of slavery and trade were increasing dissent between northern, southern, eastern and western settlers, and these tremendous conflicts were to divide the country in the most severe test of its existence — the Civil War. One of the most crucial events was the admission, in 1842, of Maine and Missouri simultaneously as states, in the famous Missouri Compromise, of which Thomas Jefferson made his comparison to a fire alarm.

In military and naval conflicts of these three wars fire became a major and frequent weapon of tactics and strategy, sometimes significant enough to change the results of confrontation and engagement, thus influencing the course of history.

In the War of 1812, the American and British burned and plundered any settlements they captured temporarily. At sea it was common practice for ships that won a battle to loot, burn, then sink ships they had conquered, unless such ships were

to be taken to port as "prizes". The American ships, although outnumbered about twenty to one by the British fleet, had greater firepower in gunnery and usually were faster and more maneuverable.[12]

When war was finally declared by President James Madison, on June 19, 1812, England was already heavily involved in military combat with France, then ruled by the Emperor Napoleon Bonaparte. In America, outright support for the U. S. war with Britain came from the western and southern states, while eastern interests strongly indicated that a diplomatic solution was best. Nevertheless, war was declared, but it developed slowly. First clashes were between American and Canadian troops along the St. Lawrence River in the vicinity of Detroit and Lake Erie, but Canadian opposition was strong and U. S. military forces were spread thinly along the frontier posts.[13]

In 1813, an American fleet under command of Captain Isaac Chauncey arrived at York, later named Toronto, defeated the military garrison, captured many arms and munitions, then burned the city before withdrawing. The British by this time were beginning to establish a blockade with seventy-two ships along the American eastern coast ports from Maine to Virginia and, since their war with Napoleon was ending, they were diverting more military troops to North America. Angered by the burning of York, the British planned to retaliate with attacks on Washington, Baltimore and New Orleans, so in 1814 transports landed in Chesapeake Bay and an army with about 4,000 British soldiers marched overland and captured Washington with ease. Some 5,000 American militia abandoned the city almost in panic.

Under command of Major General Robert Ross and Major Admiral Sir George Cockburn, the troops set fire to the Capitol, to the Treasury Building, to buildings used by the State and War Departments and to a government arsenal.

Previously, Britain had received some assistance from Loyalist citizens. On the night of September 1, 1812, twenty fires were started at one time in different parts of the city. The weather was very stormy with a high wind blowing but local citizens rallied and were able to control and extinguish the fires. The government promptly appointed special watchmen and the following night a man was discovered setting fire to a stable. He was arrested and on the subsequent day, while on his way to his trial, he was taken by a mob and hanged from a tree. This ruthless example helped to cut down on the incidence of arson as the war developed.

During the invasion of the capital city, another force of British troops went up the Pawtuxent River to attack Baltimore and Havre de Grace, but by this time American militia rallied to protect Baltimore, so the British withdrew, pausing long enough to devastate Havre de Grace with

rockets fired from ships that set many buildings ablaze.

This war brought increased use of fire rafts or fire ships. These craft were loaded with combustibles, sailed as close as possible to the enemy's ships, and, after trains of powder were poured, a flaming torch was thrown into the hold of the fire ship, blazed up quickly and raised an immediate threat to the enemy vessel which usually was heavily loaded with gunpowder.

The technique was not always successful. In 1812, fire rafts and fire ships were used on the Delaware River as part of the state flotilla of the Pennsylvania Navy. When the British were trying to move obstructions from the river at Billingsport, the Americans sent nine fire rafts down the river but the British were able to hook them with grapnels and take them ashore.[14]

In that same incident, when the British were trying to attack Fort Mifflin, the *HMS Augusta* with its 64 guns became grounded on the shallows of the river, caught fire and exploded. It was the largest British vessel in action in the Revolutionary and 1812 wars. Just behind the *Augusta* was the sloop-of-war *Merlin* which also ran aground. It was set on fire by its own crew and abandoned.

In 1814 the British were not successful in their plans to attack New Orleans. They suffered great losses when they finally engaged American troops under the command of Andrew Jackson. That savage battle proved unnecessary, because two weeks previously a peace treaty had been signed by both countries but the news did not reach the military commanders until too late.

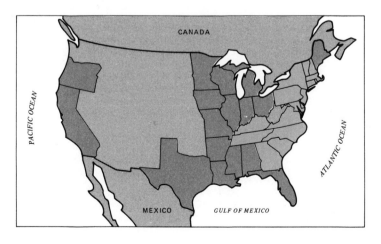

From 1800 to the start of the Civil War, the following states joined the Union: Ohio, 1803; Louisiana, 1812; Indiana, 1816; Mississippi, 1817; Illinois, 1818; Alabama, 1819; Maine, 1820; Missouri, 1821; Arkansas, 1836; Michigan, 1837; Florida, 1845; Texas, 1845; Iowa, 1846; Wisconsin, 1848; California, 1850; Minnesota, 1858; Oregon, 1859.

"Fire Engine Race" was painted (Artist unknown) in the mid 19th century and shows the commotion that accompanied movement of hand drawn apparatus to a fire. (Smithsonian Institution.)

By the 1850s some volunteer firemen wore tall hats, colorful shirts and pants, and carried unique lanterns for identity on the fireground. (Smithsonian Institution.)

PLATE CCCCLXXI

Sellers & Pennock's improved Fire Apparatus

In Philadelphia, the firm of Sellers and Pennock became well known for development of fire hose, handtubs, and other equipment. Here is a sketch of that era showing the use of a handtub, buckets, hydrant, and supply hose. (Smithsonian Institution.)

The Volunteers

In 1803, corporate authorities of Georgetown (the District of Columbia in 1871) took ownership of the engine and buckets previously purchased by subscription. Two years later the town was divided into two wards, with a fire company organized for each ward and staff officers appointed by the mayor. Every proprietor of a house was to provide and keep in repair leather fire buckets of 2½-gallon size, to be numbered and entered with the clerk of the corporation. The mayor was directed to appoint four persons annually in each ward as staff men, whose duty was to preserve order at fires and to direct where water should be obtained. These men had full power and authority to seize and compel every person to assist in extinguishing a fire. If a citizen refused, a fine of five dollars could be levied for each offense.

A year later fifty dollars was appropriated for purchase of fire hooks, axes and ladders to be kept near the engine house, and then in 1807 the mayor appointed a chimney sweep to clean all chimneys in town, similar to the fire prevention activities initiated by other colonies. All chimney flues were required to be fourteen inches square. In those days the town did not have water mains but did have cisterns and pumps by which fire buckets could be filled by hand. In 1800, the government appropriated money for two manual pumping engines, and then for two more in 1804.[15]

In those years, the area that was to become the District of Columbia was served by a number of very active volunteer fire companies which later became the nucleus for the paid Washington D.C. Fire Department. The Anacostia, Potomac and Vigilant were the first of these, all organized before 1819. They were followed by the Mechanical, the Eagle, the Columbia, the Franklin, the Perseverance, Northern Liberty, and Western Hose, and the Metropolitan and American Hook and Ladder Companies. Altogether, they represented a working force of about 150 men.

Richmond, Virginia, was another of the early cities to develop volunteer companies. In 1816, the "Richmond Fire Society" was organized for the purpose of "rendering mutual assistance in the hour of peril, and to extend the influence of effect of friendship." Each member was required to pay dues, to pay seventy-five cents for a book of regulations and to provide himself with "two buckets, two bags and a bed-socket key." The latter was for the purpose of unbolting the heavy bed frames in dwellings to remove them from the danger of fire. Membership in the Society was limited to thirty persons.

By 1819 there were three volunteer fire companies in the city, each equipped with hoses, reels and side-braked, hand-pumped engines. By 1830 the city had appropriated money for a water supply system, reservoir, water power pumphouse and private hydrants.[15]

Volunteers in New York City [16]

Fires were extremely frequent in all American cities in the 19th century, and especially so in New York. Foreign visitors were amazed to learn that two or three fires a night were regarded as quite ordinary. They had only to ascend to the roof of their hotel to see the glow of nearby blazes.

The frequency of fires was attributed to a number of causes. Among these were drunkenness and the discarding of half-smoked cigars. But arson was also widely prevalent, sometimes on the part of discontented emigrant employees, more often by businessmen intent on collecting insurance. In rural areas it was customary for quarreling farmers to threaten to burn down each other's barns. In such cases each farmer would sit up nightly for months thereafter to guard his property.

Though fires caused little loss of life, property damage was very high. The Great Fire of 1835 burned down over fifty acres in the heart of New York. Harriet Martineau, a prominent Englishwoman who witnessed the fire, was impressed not so much by the blaze itself as by the fact that even before the embers had cooled, merchants were rushing plans for bigger and better shops to replace those destroyed. To her, this typified the "go-ahead" spirit of America.

The rates and premiums of the New York insurance companies were as high or higher than anywhere in the world. Yet despite this, so many claims were made upon them that they were constantly failing.

Fire fighting in New York, though not particularly efficient, was very picturesque. The firemen, who numbered about 300 to 400 men in all, were volunteers. They provided their own uniforms, and received no compensation other than the esteem of their fellow citizens. The usual period of enlistment was seven years.

The location of the fire was indicated by a crimson ball or lantern suspended from the cupola of City Hall. The position of the ball to the cupola indicated the direction of the fire. At the same time a heavy alarm bell was rung whose strokes pounded out the ward in which the fire was located. This auditory alarm was relayed throughout the city by churchbells. One night fires were discovered in three separate locations. The lantern was shifted from one direction to another with the result that the firemen raced around in circles while the flames raged unchecked. Finally messengers were sent out to direct them, but not until fears had been raised for the safety of the entire city.

Immediately after the alarm was given, the volunteer firemen would rush to the engine houses for their apparatus and race to the fire. It was established practice for the first brigade reaching the scene to be in command. Its director became the director-general for the evening. He had absolute control over not only his own company, but also all others which arrived later. This led to keen competition.

Although a company of 20 to 100 men was attached to each engine, the engine was started rolling from the engine house as soon as the first three or four men had responded to the alarm. The engines were handsome, but very small and light. Despite their diminutive size, they were acknowledged to be superior to the larger engines that were used in Europe at that time. Americans were proud of their fire engines, and boasted that they were the best in the world. Since no horses were yet employed, the people in the streets assisted in dragging the engine with ropes.

Competition to be first at the fire was so great that often the firemen would put on only their boots and overcoats, carrying their clothes on their arms, and finishing their dressing at the fire. The firemen ran through the streets dragging their engine behind them. Some blew horns, others rang the large bells attached to the engine. At night, firemen with flaming torches dashed ahead to light the way. The sight of a fire company responding to a nocturnal alarm was awesome and terrifying, a scene that strangers to New York long remembered.

Should two companies arrive at the same time, a bloody battle quite frequently ensued between them (pictured above). Meanwhile the fire raged unchecked. After the fire, or in the event of a false alarm, the firemen adjourned to a tavern and made a night of it.

"Take up!"—"Man your Rope!" is the title of this drawing by Nathaniel Currier in "The Life of a Fireman" series. This represents action of the volunteers in New York City in 1854. (Library of Congress Photo.)

...and in other cities...

This Currier and Ives illustration shows Nathaniel Currier posed as "The American Fireman," circa 1858. (Library of Congress Photo.)

Philadelphia was another city in which competition between volunteers rose to murderous actions. Originally, the fireground struggles between fire companies were known as "pacing" contests — rivalry to secure the plugs (hydrants) nearest the burning building and disputes about which engine company was entitled to the service of certain hose companies. If an engine company arriving on the fireground could get water quickly from a friendly hose company, the pump would go into service. If the hose company did not want to furnish the water, members of the pump company would use force to try to obtain cooperation. Thus, minor skirmishes and fist fights developed.

The violence grew more serious as the incidents multiplied. One of the biggest rivalries was between the Fairmount and Goodwill engine companies, the former comprised of butchers from the Spring Garden districts, the latter manned by bricklayers from the western section of the city. Frequently,

"The Metropolitan System" in New York City began when the fire department changed from volunteer to paid. This is a Currier and Ives depiction of action during an 1866 fire. (Library of Congress Photo.)

"Facing the Enemy" is the title of this Currier and Ives lithograph in "The American Fireman" series, circa 1858. (Library of Congress Photo.)

after a fire, these two groups had major battles, with bricks, clubs, spanner wrenches and other implements used freely. Sometimes injuries were bad enough to cause hospitalization.

The worst feud in Philadelphia developed among the Moyamensing, the Franklin, and the Shiffler Hose Companies and it lasted for years. It was not uncommon for one or more volunteers to be killed in these fights, with stabbings, cut heads, and countless other injuries dropping members of these rival companies. Members of the Moyamensing group became known as the "Killers" or the "Moyas." "The history of the Moyamensing Hose Company ought to be written in blood read by the light of burning homes kindled by the torch of the incendiary," were the descriptive words of one journalist. Among crimes charged to this fire company were the shooting of a policeman who was trying to extinguish a bonfire that had been started to decoy a rival hose company, and the murders of

Savannah, Georgia

Savannah, Georgia purchased a hand engine in 1759 and fifteen townsmen organized a fire company agreeing "to keep the engine in good repair and attend upon any accident of fire". In the same year wardens and vestrymen of Christ Church were authorized to procure fifty leather buckets and fifteen fire hooks, to be paid for by a "tax upon the citizens in proportion to the number of hearths in house". Thus the first fire company and first ordinance were established in Savannah.

Louisville, KY

In 1798 the General Assembly of the state enacted a law allowing formation of five fire companies in Louisville. This required that all names of members be recorded in county court records and restricted the volunteer membership to forty men per company. From this date until 1832 the city was protected by the volunteer fire brigade which included the active athletic type of men, well-trained and operating under strict discipline. General fireground practice was to form a double row of these members to pass buckets of water to and from the fire. In severe fire circumstances sometimes women were pressed into this type of service.

four members of the Goodwill Company on Race Street, after a fight between the white "Killers" and black citizens on election night. The Moyas went through the streets, beating, maiming and stabbing all blacks they encountered. The nearby California House tavern was set afire and when the Goodwill Hose Company responded, a battle developed with Goodwill and the blacks on one side, against the rioting Moyas. The fight lasted all night while the tavern burned to destruction. Later the Moyas were charged with killing a professor of a Negro high school, killing a member of the Franklin Hose Company, and engaging in a savage gun battle with that same company.[17]

There were many other volunteer fire companies in the city, but their rivalries were milder, usually fist fights and overtuning of each other's pumpers or hose wagons. The best of the volunteer groups had excellent reputations and lasted in service for many years.

News item — 1856

**The Murder of Wm. McIntyre —
Continuation of the Coroner's Inquest**

The inquest in the case of Wm. McIntyre, who was murdered on Sunday morning last, during an alarm of fire, was resumed on Saturday night, at Long's Hotel, Third street below German.

Dr. S. P. Brown was sworn, and testified that he had made a post mortem examination on the body of the deceased. Dr. B. said —

On the left side of the thorax or chest, between the second and third ribs, and just at the end of the sternum, I saw a gunshot wound. The ball had passed inward, through the upper part of the right auricle of the heart, through the oesophagus, struck the vertebrae, deflected to the right, passed through the base of the right lung, and fractured the seventh rib, and was found in the muscle, beneath the skin, just inside the right shoulder blade. There was a contused wound on the right frontal eminence. I do not know whether it was caused by a fall or a blow. There was an old wound on one of the legs. It was not received that night.

William McMullin, sworn. — On this night week, in the neighborhood of 12 o'clock, I was standing at the corner of Seventh and South streets, talking to several persons; I heard an alarm of fire; there was a bell ringing; it rang for some time; two hose carriages passed out South street, racing; they were pretty close together; Mr. Mooney, (one of my companions) started to run out after them; I stood there a moment longer, and then started myself out South street, towards Eighth; got to the southwest corner of Eighth and South, and there met Thomas Dornan, standing there; I asked him what carriages had gone out; he said he did not know; I then heard the Shiffler coming up South street; I said, "Here comes the Shiffler." When they were near Eighth I crossed to the southeast corner, and stood leaning up against the post; they went past hollering and hooting, "Go in, Rip Raps!" "Plug Uglies." After they had passed me, I went to go across again to the southwest corner; a parcel of bricks came and fell in the street. I then crossed to the north side of South street, and went out South to see if I could not get some officer to have them arrested. When I got between Eighth and Ninth streets, I saw Constable Drew on the south side of the street. I told him to arrest; he said they were all around, and he couldn't. I got near Ninth street, and saw Officer Ryan, and I told him to arrest. I crossed to the north side at Ninth street. Just as I got on the pavement and turned go out South street, I saw four or five men behind the carriage; I saw the flash of a pistol fired from the crowd behind the carriage. I saw McIntyre; he was dead. I saw the man who had fired the pistol. I have no doubt it was English . . .

(continued next page)

News item — cont'd.

George W. Powell was next called and sworn. — Live at the southeast corner of Ninth and South streets. On Saturday evening I was in my yard and heard an alarm of fire; after 12 o'clock I saw some companies pass up; I saw two men going up ahead of one of the companies, at the corner of Ninth and South; one asked the other for the horn; he called through it, "lay it, it, Shiffler Hose;" the company was coming along quietly at the time; there were some men behind the carriage, one man dressed in dark clothes and with a Kossuth hat on, hollered "halt boys!" there seemed to be some confusion, during which the man in the Kossuth hat caught hold of a man on the sidewalk and commenced beating him over the head; I heard several persons calling out to shoot; I thought it time to get out of the way, and I went in doors; in a short time I heard a pistol shot, and heard bricks thrown; they came up against my house; the company appeared to be moving along all this time; I did not see the pistol fired; as soon as I saw the body I was satisfied McIntyre was the person who was beating the man over the head; I knew McIntyre by sight before the occurrence; I have no doubt McIntyre was shot during the difficulty.

Ephraim Garton, sworn — Lives at No. 62 Marshall street; I was going up Fourth street; when I got near the corner of Ninth, a crowd came up Ninth street and hallowed "after them boys!" — "Give it to them!" Just as the crowd hallowed the pistol was fired; the Shiffler must have been five or six pavements ahead at the time the pistol was fired; I am not a member of the Shiffler; I thought at the time that the pistol was fired by the attacking party. Before the attack was made on the Shiffler I heard no noise nor cries of any kind.

John Drew, Constable, was recalled and sworn — I saw Wm. McMullin on the night of the murder; some person cried to me to make an arrest; I do not know whether it was McMullin or not; I do not know if he went up to Eleventh street with me; I cannot recognize the man who shot a pistol; I was acquainted with McIntyre; I saw him that evening; I might have had some conversation with him; I am in conversation with so many persons every evening that I cannot tell; I do not frequently go to fires; I cannot exactly tell when the pistol was fired; I think it was about five minutes after twelve; I do not know what side of the street the pistol was fired.

The jury then retired at 8¼ o'clock, and were locked in the room, the Coroner being determined that they should come to some decision.

In about an hour and a half the jury returned, and rendered a verdict that the deceased came to his death by a gun or pistol, fired by some person to the jury unknown.

North American and United States Gazette, October 20, 1856.

Famous Volunteers

Throughout the history of the United States, but especially in the 18th and 19th centuries most political and military leaders were members of volunteer fire departments at some time. George Washington, Thomas Jefferson, Benjamin Franklin, Alexander Hamilton, John Jay, Benedict Arnold and hundreds of other famous persons in history served as volunteer firemen.

Benedict Arnold, after a highly respected military career on the American side, became a traitor when he deserted to the British army. On April 25, 1781, as a British brigadier general, he led a regiment of infantry and fifty horse troops into Richmond, Virginia where he ordered the burning of the city's foundry, powder magazine, boring mill, and two other houses. He then retired for the night to nearby Westhans. The next morning he burned more public and private buildings, muskets and munitions. (*Thereby becoming a turncoat to the volunteer fire service.* — Ed.)[22]

The Los Angeles City Fire Department first organized a volunteer unit in September 1871. It purchased an Amoskeag pumper and a hose cart which were drawn by hand. The Amoskeag was shipped from Massachusetts to San Francisco by rail, then had to be transported to Los Angeles in wagons drawn by mules.

In 1874, when the volunteer engine company asked the city council to purchase horses for pulling this heavy apparatus, the request was denied and the company disbanded.[18]

In Ohio, on December 28, 1788, Losantiville was founded, later to be named Cincinnati. By 1800, there were only about 150 buildings in this midwestern community, some of logs, others of frame and brick, and there were less than 800 inhabitants. Cincinnati did not receive its first village charter until 1802 but on July 7 of that year the council passed an ordinance establishing a volunteer fire department. All male members of a family, who were physically able, were required to respond to fires. Equipment consisted of leather buckets, fire axes and ladders thirty feet long. The fire alarm was sounded by the blowing of fox horns.

In July 1808, the Cincinnati Fire Bucket Company was organized. The first apparatus consisted of a large willow basket on a four-wheeled truck ten feet long and six feet high. The basket carried leather buckets.

Other great paintings of 19th century fireground action were made by Thomas and Wylie. This is a lithograph of a scene in 1895. (Library of Congress Photo.)

Great rivalry developed between the eighteen volunteer fire companies which had about 100 members in each company and the fire guards. There were fourteen hand pump companies, one hook and ladder company, one protection company, a number of fire wardens and a company of fire guards. One of the worst brawls occurred in 1851 when a planing mill building was burning. As the fire proceeded, ten of the volunteer fire companies were drawn into a massive fight, while the building burned to the ground. The city's mayor arrived at the scene and tried to get the men to stop fighting but his pleas were not heeded and the fight continued until daylight.[19]

It was not long before the people in Cincinnati reacted against the brawling volunteers, as had happened in other cities where safety of the public was being jeopardized by the feuds of the fire companies. On April 1, 1853 Cincinnati formed the first paid fire department in the United States and the first to use steam fire engines. Miles Greenwood was appointed chief, Engine 3 was organized as a Steam Company and Ladder Company 1 was formed. From then on, the department expanded continually as a paid organization.

The first volunteer fire company in Milwaukee, Wisconsin, was formed in 1837, nine years before the city became incorporated. This was a hook and ladder company with all members serving on a volunteer basis. The volunteer fire department continued until November, 1861 when the members were placed on a half-pay basis, serving in their civilian occupations during the daytime and on duty at the fire station at night. Milwaukee received its first steam fire engine in November 1861.[20]

In 1811, the fire department of the town of St. Louis, Missouri, was formed by an ordinance passed by the Board of Trustees. This stipulated the service required of each inhabitant, the time for exercise, and the amount of fines for persons who failed to appear for emergencies. The ordinance had simple, but direct, language: "Immediately on notice that a fire hath broke out, assemble on the spot and render the most effective service. Each and every householder, occupier of a store, or shop within limits of the town of St. Louis, furnish himself or herself with two leather buckets to be made use of in case of fire."

Volunteer fire companies were organized in 1821 and St. Louis became a fully paid steam fire department on September 14, 1857, at which time it had one horse drawn pumper and hand pulled pumpers.[21]

Flavor of the Times: On these pages extracts from newspaper columns of the mid-19th century are included to reflect the emotional turmoil of that period when Philadelphia was about to end the volunteer fire service and hire a paid fire department. The news copy is worthy of note: In those days there were few typewriters (U.S. patent granted, 1829); shorthand was just beginning to be used (Pitman, 1837); printer's type was mostly hand set (linotype, 1884); and there was no photography, no tape recorders, or other easy means of documentation. Yet, news columns carried a wealth of detail, with sufficient editorial impact to assure a faithful audience. The type, by the way, was smaller and closer set than shown here, but Ben Franklin had invented bifocal glasses in 1784, and there was time for leisurely reading.

Problems in Philadelphia — 1855

Handsome hand fire engine of Weccacoe Fire Company in Philadelphia, instituted 1800, incorporated 1833. Note ornate decorating and top hat, cape, and greatcoat of member at right. (Smithsonian Institute.)

Steam fire engine of Hibernia No. 1 in Philadelphia was built by Reaney and Neafie and was in service in 1859. (From Harold Walker collection.)

The Fire Department of Philadelphia

Mr. Editor: The municipal legislators of our beautiful metropolis have, to the surprise of our law-abiding and order-loving citizens, who are pained at its character, just passed an ordinance creating a new fire department. No doubt the large majority of both branches of the city legislature who voted for this bill, were actuated by the best motives, but they have committed a sad and fatal error. The re-organization of this important department, which their action contemplates, is, to give it the mildest designation, a mere patching up of a bad system.

The volunteer fire department of Philadelphia is full to overflowing of discordant, jarring and belligerent elements, while its principal and ascendant material is mischievous and lawless in the extreme. There is no longer any denying of these too palpable facts. Thus constituted, then, and abounding in such ingredients, *it cannot be made to govern itself*, and all attempts to accomplish that end will prove woful failures. The reform ought to have been radical and thorough, and nothing short of complete and entire municipal control should have been, for a single moment, listened to by the city fathers.

The ruffianly and vandal scenes enacted in different parts of the city since New Year's day, have shown, beyond all question, that the volunteer firemen are the open and bold defyers of the law and its officers, and that they cannot be kept in subjection except at the muzzle of the revolver, and then only when the weapon is wielded by resolute hands, sustained by brave hearts.

The dark, bloody and dreadful deeds of the firemen prior to the creation of the Marshal's police, have not yet been forgotten by our citizens. That body of men was scarcely organized ere the fire department — the various companies making common cause — arrayed itself in the most hostile attitude against the new police force. This rebellious deportment was not confined to the so-called "rowdy companies," but was a rising *en masse* — companies claiming the highest position of re-

spectability in the department taking the lead in this opposition to all legal authority. From that day to the advent of consolidation it was a contest between the firemen and the police authorities for the supremacy. The law, it is true, was triumphant, yet in maintaining its ascendancy the officers were often forced into the most hazardous and trying situations. In more than one instance they were compelled to shoot a rioter dead, or yield up their own life as a sacrifice to the spirit of ruffianism with which they were contending.

Twice did the firemen invoke the State Legislature to allow them to do as they pleased; but here a deaf ear was very properly turned to their presumptuous and unreasonable appeals. Our new Councils, though, have been more complaisant. They have permitted these men to coolly dictate their own terms, and given them all they asked. They have virtually made them their (the Councilmen's) own masters, and what was, above everything, most desired by them (the firemen) made them the masters of the police!

The first fruits of this astonishing legislation which, however well meant, is still unfortunate, have been witnessed in the disgraceful riots that began on New Year's morning, and have not yet ended.

There are about seventy engine, hose and hook and ladder companies at present in the old city proper and the built up districts. Those in the boroughs of the county, will increase the number within the limits of the consolidated city, to about eighty. According to the public records, since the month of October, 1850, the following companies have been declared out of service for being engaged in riots: Fairmount, twice; Northern Liberty, twice; Good Will, Vigilant, Hope, Assistance, Independence, Spring Garden, Weccacoe and Kensington engines; Northern Liberty, twice; Marion, Good Will, Niagara, Independence, Moyamensing, South Penn, Franklin, Lafayette, United States, Hibernia and Weccacoe hoses, and the Relief hook and ladder. The Weccacoe Hose Company having been reported to

the Court of Quarter Sessions twice in one year, was, on the second conviction, as the law directs, disbanded. One or two of the other companies in the above list have been twice put out of service; and, besides those named, two or three others incurred the same penalty, the names of which cannot be recollected with certainty. At least a dozen other companies were reported to the Court, but escaped the penalty.

The following companies, concerned in riots, have not been put out of service, although some of the affrays were of the most outrageous character: Good Intent Engine, William Penn Hose, Shield Hose, Washington Engine, Globe Engine, Carol Hose, Hand-in-Hand Engine, Pennsylvania Hose, Mechanic Engine, Eagle Hose, Philadelphia Engine, Western Engine, Franklin Engine, Western Hose, Fairmount Hose, Spring Garden Hose, America Engine, Taylor Hose, Jackson Hose, and Jefferson Hook and Ladder Company; together with three or four companies in Germantown, and other places in the rural portion of Philadelphia.

These incontrovertible statistics show that, within the short period of a little more than four years, more than one half of the companies of the whole fire department have been concerned in regular riots, which were participated in either by members, runners, or adherents; while more than one-fourth of the companies have been declared out of service.

But this is not all. A number of the fire companies claim to be more "respectable" than their other brethren in the volunteer organization. Yet can hardly half a dozen companies be mentioned, as the police records will show, which the authorities have not had some cause of complaint against, during the last four years past. Even companies that are generally quiet and orderly do not seem to be able to escape the contamination of the virus of the miserable system which is about being re-inflicted upon the community.

LAW AND ORDER

North American and United States Gazette, January 29, 1855.

The School of the Fire Plug.

In a large city, which tolerates the enormous evil of a volunteer fire department, with all its attendant clannish excitement and hostility, and debasing associations, the schools of the fire plug are numerous, and, for the communication of their poisonous lessons, very efficient. Some time since, Philadelphia was cursed with an immense number of these Jakey and Sikesy academies, and the youth of the city were threatened with a general demoralization. The Marshal's Police and the free reading rooms have greatly diminished the evil, but owing to the continuance of a radical cause, these schools of degradation — these councils of wickedness — are still numerous in some of the districts — and especially in Kensington and Penn District. There we may yet see the corner crowds of disorderly juveniles which were once noted features in the lower part of the city. The police seem to care nothing for such assemblages, although it is plainly their duty to break them up by arresting those who refuse to abandon the practice of corner lounging. We hope they will give the matter serious attention. These fire-plug circles are the nurseries of crime, for which the police are constantly called upon to make arrests. There the youth learns to bully and riot, to carry deadly weapons, to swear, to drink, and fifty other equally detestable accomplishments. The Marshal's Police can effect much; but we are not justified in hoping that these degrading assemblages will be completely destroyed until the volunteer fire department is abolished, and free reading rooms are established in all parts of the city. We know that many persons ascribe the evil to a certain restlessness of youth, which must have some active amusement. We doubt this account of the disease; but if it be a true statement, let gymnastic societies for healthful physical exercise, upon the plan of the German Turnvereins, be substituted for the fire companies, to enable the young men to expend their superfluous energy. With these exercises of the body mental culture may be readily combined. Then we are certain there would be no more schools of the fire-plug.

North American and United States Gazette, March 31, 1854.

How to Live a Hundred Years

As a Beverage, use the old Wheat Whiskey, made from the choicest wheat, and the pure mountain waters of the Keystone State. It is as mellow as a peach, fully ripe, and as smooth as oil; put up in cases of one dozen quart bottles, and labelled "Alter & Williston's Old Wheat Whiskey, 1835, unequalled in the world," and sold by Alter & Williston, at Nos. 228 and 230 South Front street.

North American and United States Gazette, August 30, 1854.

... And a Paid Fire Department is Urged — 1855

This one-paragraph editorial from a Philadelphia newspaper emphasizes the critical situation in the city in the violent days just before the Civil War. Finally, in 1870, after 120 years of service, the volunteer fire companies were replaced by a paid fire department.

The Fire Department

The city of Philadelphia is at this moment in a very awkward, and, indeed, a very perilous predicament. During the last forty-eight hours, it may truly be said to have been literally at the mercy of the elements — at least at the mercy of the winds and the flames. It is to Providence that we should feel thankful that an awful calamity has not overwhelmed us; that immense districts, in the very heart of the metropolis, are not to-day a heap of smoking and blackened ruins; that millions of dollars' worth of property is not virtually annihilated; that hundreds of lives have not been destroyed; and that a terrible sum of human distress, and suffering and destitution does not now afflict this community. We have thus far made a narrow escape — an escape from what might have been a disaster of fearful magnitude — and it is rather to the interposition of an ever-good and watchful God, than to any sufficient precautionary measures of our own, that we are indebted for the deliverance. Under the law recently enacted by Councils, a large portion of the volunteer Fire Department went out of service on Saturday night, at 12 o'clock, while that part which accepted the ordinance, and agreed to remain on duty, were not in a condition to act under the prescribed authority. Our ever-vigilant and laborious Mayor foresaw this dangerous and threatening state of affairs. He perceived that, in consequence of the injudicious and improvident action of the local legislature, which provided for the possible retirement from service of the entire body of our firemen, without making any arrangement, or supplying any means to avert the evils of such a contingency, a frightful exigency was inevitable, and with a highly commendable devotion he began immediately to repair, as well as he was able under such difficult circumstances, the indiscretion of the Councils. He obtained the sanction of high professional advice to instruct the fire companies consenting to continue in service, that they might act in any emergency, irrespective of the ordinance, and he issued orders accordingly. Having, moreover, received positive and reliable information that the disaffected and recusant fire associations were rife for a fierce and violent outbreak; that they had laid their plans and made their preparations for it; that they had, in fact, concerted measures for causing simultaneously conflagrations in opposite extremities of the city, and that murderous attacks were to be made upon the men who should proceed with their apparatus to protect the property of the citizens — the Mayor did all that it was in his power to do to counteract and defeat these formidable designs. He ordered his entire police, the night and day force, officers and men, on duty, and they were

kept afoot in the streets of the city, ready for immediate action either to prevent or suppress a disturbance. He visited a number of the engine houses of the hostile fire companies, and found the members collected there in a high state of excitement, and to all appearances, only waiting the opportunity to commence a scene of the wildest disorder and outrage; he told them that he had received intelligence of their purposes; he remonstrated against the disloyal and criminal spirit they were manifesting; he notified them of the measures he had adopted to resist and punish their conspiracy, should they attempt to execute it, and in several cases in which he deemed the act necessary, he closed the houses, and stationed policemen to guard the premises. Thus prepared to meet the menaced emergency, he had made arrangements by which, in the event of a riot, he could proceed at once, in person, to the locality of the outbreak, and take the lead in the efforts to quell it. Happily for the public the disorders, which there was so much reason to apprehend, did not occur. But can there be a just doubt that the elements of a frightful and destructive disaster were existent in our midst during the forty-eight hours which have elapsed? Suppose that a serious fire had happened any where in the thickly built sections of the city, without the agency of any incendiary, between Sunday and Monday morning, what would have been the probable result? With one half of the firemen, with their engines and hose, out of service; with the other half unorganized under the ordinance, while embarrassed by doubts as to the binding obligation of its cuppling provisions restraining the companies within defined limits of action, and threatened, moreover, with a deadly assault in case they should dare to venture forth for the suppression of a fire, what might have been — nay, what in all probability would have been, the issue? Would not the tremendous hurricane which prevailed at the time have fanned the conflagration into terrific fury; borne it with resistless and rapid sweep from house to house, and block to block, and street to street, until it had laid the city, or a vast portion of it, in ashes, and produced a devastation which would have shattered fatally the fortunes of the community, and required years for its complete reparation? Let us not deceive ourselves into supposing that, because we have escaped it, the danger did not in fact exist. It did exist. The materials for it were supplied; the spirit of insurrection was aroused and ready for the onset; the improvidence of the Councils had facilitated and apparently invited the peril; and nothing was needed but two or more accidental fires, occurring contemporaneously in different quarters of the town,

to set the torch to a train of mischief and wickedness, which in its development would, we fear, have baffled description.

Our conviction is, that we owe, under Providence, to the vigorous exertions, the timely and sharp vigilance, and the masterly dispositions of the Mayor, the rescue from a calamity which was imminent and otherwise unavertable. We know too well the desperate and rebellious character of the worst class of the firemen — we have seen too much of their insubordinate, riotous and malignant conduct, to imagine for one moment that the security we have enjoyed in this crisis, is attributable, in any measure, to their virtuous forbearance, their loyalty, their respect for the laws, and regard for the public welfare. The only considerateness, which, doubtless, actuated them in the juncture which offered to their malcontent spirit so favorable a chance of vengeance — the only motives to which the community may ascribe its protection against their vindictive and remorseless licentiousness, are those of fear in view of the powerful and well disposed means which they knew were held in readiness against any demonstrations of violence they might attempt. This was the respect, which subdued their refractory and lawless temper; this was the impediment which prevented their malicious purpose from emerging into act; this was the intimidating dread that held them at bay, and compelled that submission, whose merit, if it have any, is but the merit of necessity. Thanks to the discreet and devoted guardianship of our Chief Magistrate, to which alone the citizens are indebted for their peace and safety throughout the perilous chance of the time. But for his opportune precautions, his anxious watchfulness, his efficient exertions, we might have been obliged to record to-day, a dark and direful chapter in the history of Philadelphia.

In short, the existing situation of the department presents an exigency imperatively demanding the instant and energetic intervention of the Councils. They have produced a condition of things full of peril and embarrassment to the community, and they must decide at once on measures fit to remedy the evils which have grown directly out of their want of wisdom and resolution in dealing hitherto with this vital interest. We still believe that a paid department of sufficient strength is the only proper alternative. We are convinced that experience will yet, if it has not already, prove the fact. But, at all events, something must be done, and done speedily, to repair the present dangerous position of affairs, or the worst consequences may confidently be anticipated.

North American and United States Gazette, April 2, 1855.

When in March, 1857, the ferry-boat *New Jersey* was burned to the water's edge while making passage from Arch Street wharf to Camden, a spark having set fire to a load of hay on board, sixty-seven persons were either burned to death or drowned amid the floating cakes of ice in the river. The volunteer firemen, attracted by the light of the flames, rendered conspicuous service in rescuing from drowning or burning to death about 100 passengers. In doing this, many feats of personal heroism were exhibited; and in more than one case, the rescuers contracted illnesses which, sooner or later, caused their death. This is but a glimpse of the better side of the picture, which goes far towards condoning the offenses of the hot-headed fellows who allied themselves with the fire service only to cast disgrace upon a time-honored tradition.

Centennial History of the Philadelphia Fire Department 1971

Development of the Fire Hydrant[1]

The fire hydrant is of relatively recent origin, having been developed within the past 200 years. This fact may seem strange since fire has always been regarded as the greatest destructive agent known, and methods to combat fire have been practiced by mankind throughout the ages. There are, however, numerous and important factors that account for this seemingly slow development, and an effort will be made to trace some of the historical events leading up to present-day firefighting and hydrant methods.

Probably the greatest influences in the development of the fire hydrant were the methods used to distribute water. The early water systems transported water under low head (*i.e., pressure — Ed.*) and it was not until in the eighteenth century that street mains were constantly and sufficiently charged.

As pipe materials and jointing methods improved, pressures were increased and the method of distributing water was changed from low service to pressure service. At first, pressure service was intermittent in character and was so practiced on the Continent and in England. Under this arrangement, water service was only available at intervals, was entirely unsatisfactory, and soon gave way to constant high-pressure distribution.

Steam fire engines first made their appearance in the 1820s, but it was not until the latter half of the nineteenth century that they came into general use. Insufficient water distribution facilities may have been in part responsible for the slow progress of power-driven pumps. The use of power-driven engines was responsible in a large degree for improvement in hydrant construction.

History records that spring water was first brought into London through lead pipes and conduits in the thirteenth century. Between the years 1606 and 1620 many pipes were laid in the London area. These pipelines were made by boring out wooden logs, a method known and used for many centuries. This type of pipe material and the method of joining pipe sections together would not permit very high pressures. As a result, low-pressure service was practiced on these early systems and the practice continued until the nineteenth century.

The first effort to obtain water directly from the street main for fire purposes was very crude. First a hole was dug in the street and the water main located; then a hole was bored in the wooden main and the water was collected from the street hole either by bucket or suction hose and fed into the pump. After the fire was extinguished, a wooden plug was driven into the hole and the street surface was restored.

The use of a wooden plug in this manner was the origin of the phrase "fire plug," a term still commonly used today. The letters "F P" are still widely used to mark the location of "fire hydrants." The word hydrant derived from hydro (Greek: *hudor*, water) probably originated in the United States.

Trouble was experienced with this method in that dirt fouled the pump valves, streets were badly washed and, with the advent of street pavements, excavating the holes became more difficult.

The next improvement of "fire plug" design followed the introduction of cast-iron pipe. Although cast iron had been used for water pipes in France since the year 1664, it did not come into general use for water mains until the beginning of the nineteenth century. With the increased use of cast iron, sockets or branch fittings with openings were placed at intervals in the water mains. A later law (1847) prescribed that these fire plug sockets be inserted in the water mains at intervals not to exceed 300 feet. Wooden plugs were inserted in these openings and a metal or wooden shield for the wooden plug extended from the main to the street surface. The shield housing the plug prevented the soil under the street surface from being washed away.

To prevent excessive waste of water and also provide a more constant flow to the pump suction the next development was the canvas cistern in the year 1820.

With the increased use of cast iron, distribution pressures were stepped up, and this led to the development of the standpipe with a tapered end for insertion directly into the main, and at the opposite end, a hose connection outlet. The brigades soon discarded the canvas dams, preferring to connect the hose to the standpipe in order to utilize the main pressure.

It was soon realized that the "plug type" of hydrant had definite limitations and was practically useless under conditions of high pressure. The pulling of the plug and the insertion of the tapered standpipe against good pressure was sometimes a difficult and always a wet task. This fact led to the use of a controlling valve or cock.

In Hamburg, Germany, a connection was made to the side of the main and a stop cock inserted which was operated by a vertical rod from the street. The hydrant tube curved up to the street level where it terminated with one or more hose connections.

In England, the hydrant (fire cock) was connected to a vertical branch from the main or to an elbow attached to the main, and the whole arrangement called a ball hydrant.

A serious objection to the ball type of hydrant is the fact that when pressure is taken off the main the ball drops from its seat allowing dirt to enter the main from the hydrant box. When pressure is reestablished the water may be contaminated. To overcome the possibility of water contamination a device known as a conversion element is fitted into the hydrant tube in place of the ball. This element consists of a brass dome fitted in a guide and held upward against the seat by a strong spring.

Location of the hydrant below the ground surface (flush type) and the transporting of the standpipe or nozzle section has been the practice in England.

Soon after this country was settled, the construction of public water supplies was begun. Some of the early systems with approximate dates are as follows: Boston, Mass., 1652; Bethlehem, Penna., 1761; Providence, R. I., 1772; Morriston, N. J., 1791; Plymouth, Mass., and Philadelphia, Penna., 1797. By the year 1800 there were fewer than 20 water works in the United States. Underground cisterns were primarily used as a water source on these early systems, and in many cities remained popular after the introduction of the hydrant. In the beginning these reservoirs were either supplied by rain water, or were kept full by manual pumping. With the installation of street mains direct connections were made to service these cisterns and their use continued as a supplement to hydrant service.

The first mention of hydrants appears in a Philadelphia report of 1801. The following quotation is taken from the 1801 report to the Select and Common Council by the Committee for the introduction of wholesome water in Philadelphia.

"The plan of hydrants that were first used has been wholly rejected and another, which answers better, substituted. Of the latter thirty-seven (37) are now fixed in different parts of the City, and six (6) more are in readiness for the same use."

It is assumed "the plan of hydrants" that were formerly used and "rejected" were of the common plug type. By the year 1811 Philadelphia claimed 230 wooden hydrant pumps and 185 fire hydrants.

Cisterns were used in New York City and were generally of 100 hogshead capacity (6,300 gallons). Fire plugs were mentioned as used in the city as early as 1807 but these were most likely "plugs" and not hydrants. It is related that the first fire hydrant in New York was installed in 1817 in front of the dwelling of Mr. George B. Smith of Engine Company No. 12 on Frankfort Street.

Fire plug arrangement with canvas cistern.

In 1854 Detroit reported 65 street reservoirs in addition to hydrants; and in 1864, it had 249 hydrants, and 127 cisterns. The cisterns were connected to water mains by 3-inch lines but some were increased to 4 inches, due to insufficiency of supply, as explained in the report:

"Since the introduction of steam fire engines into the city, complaints have been made of the insufficiency of the supply of water in the street reservoirs in some localities; this may be remedied by introducing an additional stream into them."

In 1861 Louisville had no fire hydrants and 124 street cisterns. The following extract from the Louisville report of 1861 is interesting:

"One hundred of the street cisterns for the Fire Department are connected with the water pipes. No fire plugs are in use, their efficacy for supplying steam fire engines with water has heretofore been regarded as doubtful. In some of the Eastern cities they are, however, used in addition to street cisterns and regarded as equally efficacious, and much more economical in construction and the use of water. . . ."

Wooden hydrant pump, Fairmount Park, Philadelphia.

53

Taking a hydrant in New York.

In 1864 Boston had 1,530 hydrants and 96 cisterns with 15 of the cisterns connected to water mains.

Philadelphia, in 1865, had over 3,000 hydrants of the post type in the city. Philadelphia apparently strongly favored hydrants over cistern service.

Cisterns had many disadvantages including high first cost, inflexibility due to wide spacing, limited capacity, excessive water losses due to leakage, dissipation of available system pressure and the total elimination of direct streams, a preliminary service very effective in the extinguishment of many fires. Cistern service made it necessary to depend entirely on engine streams, a fact that increased the popularity of hydrant and hose service. Therefore, with the improvements in distribution practice, including high pressures, supply pipes of larger diameter, and the production of more sturdy and reliable hydrants, cisterns were gradually replaced.

It has been previously emphasized that fire hydrant development was largely dependent on methods of distribution. Many of the early systems were insufficient in pressure and capacities to accommodate present fire hydrant service methods. The deficiencies of those early water systems were recognized first by

Fire hydrant, Richmond, Virginia, 1830.

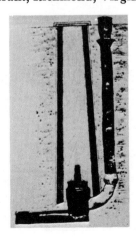

owners of factories and high business buildings where system pressures were inadequate for fire protection. Pumps were installed on these individual premises and the benefits were so apparent that it was not very long before high pressures were extended to entire water works systems.

It may be said that the United States led the way in adopting pressure service, and also in the development of the fixed-post type fire hydrant with control valve constantly charged with pressure.

The first post type hydrant in the United States was designed by Mr. Frederick Graff, Chief Engineer of the Philadelphia Water Works and was used in Philadelphia about the year 1801. It had a combination hose and faucet outlet, and was a "wet barrel" design with the valve in the top.

It is certainly not surprising to learn that the "wet barrel" design gave trouble.

An improved hydrant with the valve at the base and facilities to drain the barrel was made in 1803. These first hydrants, the forerunner of the modern post hydrant, were inserted by a tapering joint into the wooden main and were sometimes carried by the brigades, English fashion. With the introduction of cast-iron water mains, sockets or branch fittings were inserted in the line at intervals and permanently connected hydrants of the flush type became more popular.

The hydrants that followed differed in detail depending on the manufacturer, but for the purpose of a general description the hydrant and installation included the following features:

A branch line was installed from the street main to the sidewalk to which was attached a "hydrant" bend. The hydrant was connected to the bend and terminated in one or more hose outlets either below the ground level (flush type) or above the ground level (post type). The flush type hydrant required a wooden or metal housing or hydrant box.

Following the English practice the flush type of hydrant at first was the more popular. The personal whim of the fireman appears to have had considerable influence in adoption of the post type of hydrant.

19th Century News Items

To Our Firemen

We do not for a moment doubt the intrepidity, energy and watchfulness of our firemen; yet, as we have learned a serious lesson from the numerous recent disasters made by the devouring element in our neighbor cities, it would be well to be on the alert, and keep, if possible, a more vigilant eye than ever has been kept, over our fire department. It may be that there are secret incendiaries, who visit the various cities, setting fire to property, so as to enable them to carry into effect with greater facility, their fiendish purpose of plunder. The frequent fires which have so lately taken place in the north, the east and the south, might perhaps be traced to some such cause. There are men in the world black-hearted enough to do any thing, and who knows how soon Baltimore may be visited with such scoundrels. We have cause to rejoice that our monumental city has been singularly fortunate thus far, as regards fires, but it may come "as a thief in the night," and we cannot be too cautious or too well prepared to meet it.
Baltimore Clipper, October 11, 1839.

The Friendship Fire Company

Agreeably to arrangement the Friendship Fire Company left by the early train yesterday morning for Lancaster, Pa., where they will to-day join a grand firemen's procession. Before leaving the company was presented with a beautiful silk standard by Miss Holtzman, on behalf of the female friends of the company. Col. S. S. Mills received the gift and thanked the ladies for the present. The flag is about 16 feet long, and has a polished ash staff, which is pointed, with an eagle perching. It bears in its folds the name of the company in gilt, also two hands clasped in friendship. After the presentation, preceded by Linhardt's band, the company with its gallery engine moved through Baltimore, Aisquith, Gay, High, Centre and Calvert streets, to the Calvert station. The engine was then placed on an open car, and at a quarter past eight o'clock the train moved off. While on the march the bells of the Independent and Patapsco companies rung out a salute. The new equipments provided for the visit consist of black coat and pantaloons, red shirt with black enameled belt and shield bearing the numerical "3", the number of the company. The cap is of drab cloth with the same numerical in silver on the front. — There were about forty of the members went on, and they made a fine appearance as they passed through the streets. They will return on Tuesday next, when several of the fire companies will turn out to escort them to their headquarters.
The Sun, September 12, 1857.

Runaway at the Lexington Market

Yesterday morning a horse attached to a buggy belonging to a Mr. Conway, of the county, and left standing at the Lexington market, took fright at a kite flying near, and ran off at full speed. A small colored boy, in whose charge the establishment was left, was knocked down by the plunge of the animal and somewhat injured. The frightened animal proceeded several squares out Lexington street, but on attempting to turn a corner the buggy was dashed against the curbstone, overturned and broken into fragments. The animal was then secured. The affair caused considerable excitement in the market.
The Sun, September 12, 1857.

"Washing Out" in New Orleans

An exemplification of the beauties of the volunteer fire department system was recently afforded in New Orleans. Company Number Sixteen having rendered itself peculiarly notorious for its rowdyism, and destroyed the contents of the house of a colored woman, by what is known as a "washing out," the local authorities thought it time to put a stop to such proceedings. Accordingly, taking advantage of the first opportunity, Recorder Filleul fined the foreman of Company Number Ten for some misdemeanor. Indignant at such an infringement of their sovereign rights, the members and rowdy adherents of Number Ten determined to give the Recorder a "washing out," and afterwards to join forces with Number Sixteen, in order to "wash out" the Parish Prison and liberate some of the firemen who had been incarcerated there for participation in the destruction of the colored woman's house. This "washing out" process consists in the engines being placed in front of the doomed premises, and then dashing the water, with all the force of the apparatus, against windows, doors, etc. Of course not a whole pane of glass remains in any sash after such a display, and all the furniture of a house is ruined by the water. Not being quite submissive enough to stand this, the Recorder had the whole police force, available for the occasion, collected at the second district station, and armed with bludgeons. Number Ten mustered one hundred and sixty desperadoes, who, headed by a band of music, were preparing to march to the Recorder's residence, when they heard of the reception awaiting them. The musicians backed out, and the rowdies slinked away. But, about an hour after midnight, they collected again, and thinking that the police had been disbanded for the night, reorganized, and, preceded by a drum and fife, marched up Conde street, to join the forces of Number Sixteen. Fortunately, they were mistaken in their calculation; for on reaching Jackson square they were met by a strong body of police, who arrested some eighteen or twenty of them, and placed them in the calaboose. Number Sixteen, being on the way to join Number Ten, heard of the arrest and incontinently beat a retreat. Now is not this a pretty state of affairs to take place in a city of one hundred and twenty thousand inhabitants! Here were two companies, incorporated ostensibly for the extinguishment of fires, undertaking to break open the city prison and liberate offenders, and to mob the residence and destroy the furniture of a public officer, for the mere performance of his duty. We in Philadelphia have not yet reached such a depth of degradation as this, though we are badly enough off in all conscience. With the material we have in our "fire department," there is no telling how soon we may descend to it.

Wherever the volunteer system extends, its evil influences may be seen. In New Orleans it undertakes to mob the authorities, destroy their furniture, and break open the prison. In Philadelphia the firemen murder each other, shoot the officers of justice, nullify all laws made to regulate them, and hold a rod of terror over legislators and councilmen, so that, however much the people may demand reform, their influence is sufficient to defeat it. In Newark they refused in a body to do duty, leaving property entirely unprotected, without a moment's warning, because of their displeasure at the acts of the city government. In Cincinnati they became so ungovernable and riotous that it was found absolutely necessary to disband the whole department and establish an entire new force composed of men paid for their labor. In Boston and Providence they have precipitated similar changes, and the City Councils of New York are now considering the propriety of establishing a paid fire department. With these facts before them, why do our own authorities hesitate to accomplish a reform which must take place sooner or later?
North American and United States Gazette, August 30, 1854.

Trial of a Hand Engine

The Kensington engine was again tried on Monday evening, with a view to compare her powers with those of the steam fire engine, "Young America." Eight hundred and fifty feet of hose were attached, and through a ⅞ inch nozzle, a stream was thrown a distance of 90 feet. One thousand and fifty feet of hose were then attached, and through the same nozzle, a stream was thrown to the distance of 80 feet. This is remarkable playing for a hand engine. But what is the use of comparing it with that great performance of the "Young America," lately witnessed. The same body of water was not thrown, nor could the men on the engine, with their utmost expense of muscular powers, maintain the stream for over a few minutes. The "Young America" could play the same stream for hours, and there would be no expense of human exertion. *North American and United States Gazette, June 13, 1855.*

Illustration of the U. S. flag ship Hartford attacked by the Confederate ram Manassas and a fire raft in the Mississippi River just before Union forces isolated New Orleans. Action occurred April 24, 1862 when Hartford was passing Forts St. Philip and Jackson. (U. S. Navy Photo.)

1861-1865

Chapter III

"... you I propose to move against (General) John-ston's army, to break it up and get into the interior of the enemy's country as far as you can, inflicting all the damage you can against the war resources..."

Lieutenant General U. S. Grant to
Major General W. T. Sherman[1]

The Civil War, otherwise named the War of the Rebellion, or the War Between the States, was a spasm of division and turmoil that threatened to dissolve the United States of America completely. Without question, it was one of the most strenuous, emotional periods in the country's history, even though it was not total war on the scale of World War I or II. The new nation, less than a century old, was heavily engaged in this military disaster that pitted brother against brother, family against family, and neighbor against neighbor, but continued its normal political processes, including expansion of business enterprise and development of new lands. Three new states were admitted to the Union, and a remarkable number of new inventions and technological developments were stimulating the growth of industry, cities and states.

Fire in the Civil War

In the four-year period of the Civil War there was the normal incidence of major fires in larger cities, but in the broad-scale movement of armies, fire was used flagrantly by northern and southern military commanders. The effect upon the civilian population, north and south, was a tremendous personal shock, because these burnings resulted in devastation of business enterprise and personal property, for the sake of maintaining military advantage. In contrast, many civilians, on both sides, deliberately destroyed their own homes and possessions, rather than allow them to be used by invading forces.

Before the Civil War, fire had been used violently by individuals and groups responding to the political crises of the time, mostly the slavery issue or the tariff. But these were small scale incidents; it remained for the military commanders to use fire as a weapon of war that would stagger the imagination and hopes of the civilian populace.

When Lieutenant General Ulysses S. Grant assumed command of the northern armies in 1864 after appointment by President Abraham Lincoln, not much time passed before he explained his personal strategy for bringing the war to conclusion. In essence, Grant decided on a campaign of annihilation of the southern military force and a staggering campaign of terror for the civilian population. This strategy was identified quite clearly in his orders to his field commanders, particularly to Major General William Tecumseh Sherman and Major General Phillip Sheridan. Initially, General Grant began to bring his armies into a cohesive group in the vicinity of Washington, DC, and Pennsylvania, to solidify the distraught Army of the Potomac, but at the same time he was issuing

Just before the end of the Civil War a spy ignited a major fire in Barnum's Museum in New York City and also confessed to starting five hotel fires. For these actions he was hung. (See page 67.)

orders for mobile forces to burn and otherwise destroy the areas where southern forces would move.[2]

A few months before, the Confederate General Jubal Early had conducted a remarkable form of guerrilla warfare in the lower Shenandoah Valley and, with a force of about 15,000 men, he threatened the Potomac Border and forced the Federal troops to divert 40,000 soldiers to combat his efforts. He sent a cavalry column into Pennsylvania, burning Chambersburg in reprisal for the towns that the Federal General Hunter had burned in Virginia. General Grant could not tolerate this type of aggravation for his military objectives. He called upon Major General Philip Sheridan.

Much historical attention has been given to the march of General Sherman's armies through Tennessee, Georgia and back into the Carolinas, and to his deliberate destruction of the cities of Atlanta, and Savannah, Georgia. But the actions of forces under General Phillip Sheridan in the Shenandoah Valley created similar destruction on a massive scale. In 1864, as part of his plan of strategy, Grant gave Sheridan specific orders to destroy all resources in the Valley that might possibly be of some usefulness to Confederate forces. The effect of this, Grant reasoned, would be to cripple mobility of the Southern columns, whose supply lines would be in-

terrupted, or extended drastically, thus hampering the fast-strike capability of the Confederate cavalry and mobile infantry. He asked for the Valley to be leveled to a "barren waste," to be devastated so it could not be used by General Early or any other southern force as a military highway for granary. Sheridan responded with alacrity, stating that his troops had destroyed more than 2,000 barns filled with wheat, hay and farming equipment, more than seventy grain mills and thousands of cattle and sheep in the valley.[3]

General Sheridan was not only obedient to the orders of his commander, he was a brilliant tactician and a strategist who drove his troops by the sheer force of his energy and leadership. When things went wrong he could explode in fury, such as the time when one of his aides was found in a field with his throat cut, apparently by the southern guerrilla troops. In anger, Sheridan ordered every house, barn, and outbuilding within five miles burned to the ground. A short time later, as his army withdrew toward the north, he ordered the upper Shenandoah Valley laid waste. A journalist traveling with the troops, reported that the sky "was black with smoke of 100 conflagrations, and at night a gleam brighter and more lurid than sunset has shot from every verge." Sheridan's orders were to burn no dwellings, but if a burning barn

58

"We have devoured the land . . . to realize what war is, one should follow our tracks."

General William T. Sherman in a letter to his wife.

Two members of the First Regiment of Fire Zouaves in their picturesque uniform which consisted of a fireman's red shirt, gray jacket, and gray flowing trousers tucked into leggings or boot tops. Two hundred Fire Zouaves were killed, wounded or taken prisoner during the Battle of Bull Run. The Zouaves were all volunteer firemen and, as the Eleventh New York Volunteers, were among the first regiments from the city to respond to President Lincoln's call for militia.

happened to be close to a house the latter usually was destroyed by flame. The correspondent mentioned that the devastation "carefully illustrates the horrible barbarity of war."[4]

Initially, General Grant wanted to apply his strategy of annihilation just to the Confederate military forces, and he issued orders that civilians living in their homes were to be allowed to keep their property in safety. Pillaging by northern military troops was discouraged and punished, but otherwise every effort was made to prevent any food, forage or useful material from falling into the hands of the mobile southern forces.

In southern states, General Sherman, operating within the scope of Grant's military orders, had a different purpose. Much of the aggressive movement of his troops was aimed at destroying the enemy's resources, but Sherman also wanted to strike terror into the hearts and minds of the civilian populace of the south. He believed that if

the enemy people experienced the complete shocking terror of destruction in war, they would lose enthusiasm for supporting the southern cause and, when the southern armies lost the support of their people, their military efforts would collapse. General Sherman was quoted as saying ". . . they must feel the effects of war . . . they must feel its inexorable necessities, before they can realize the pleasures and amenities of peace. If the people raise a howl against my barbarity and cruelty, I will answer that war is war, and not popularity-seeking. If they want peace, they and their relatives must stop the war."

Another of his quotations stated, "War is simply power unrestrained by constitution or compact. You cannot qualify war in harsher terms than I will. War is cruelty, and you cannot refine it." Military records and historical reports offer ample testimony of the tremendous destruction created by the forces under General Sherman.[5]

59

An eyewitness report of the devastating march to Atlanta appeared in *Harper's Magazine* of July, 1865. This quoted a diary of Brevet Major George Ward Nichols who served on the command staff as aide-de-camp to Sherman. Major Nichols called attention to the problems of leading an army on such a complicated march which traveled from Rome, in the extreme northwestern corner of Georgia, near the border of Alabama, southeastward to Savannah. Then, after occupying that burning city, the army turned north and moved forward 150 miles to Columbia, South Carolina. Then it moved northeastward 200 miles to Goldsborough, North Carolina, after which it went northwestward to Raleigh and Chapel Hill in the same state.

The army moved in the classic advance of foot troops — in two main columns, two corps in each column, spanning a width of forty to sixty miles as it advanced. Following the orders of Grant and Sherman, the troops destroyed everything of value in the corridor between Kingston and Atlanta. On January 27, Savannah burned in a tremendous conflagration, which included the shattering explosion of the Confederate arsenal in that city. One report stated that there was no organized fire department action as the city burned, and indicated this as an important limitation in the city's defense.[6] A month later, the army columns were in South Carolina. Charleston was evacuated by the Southern forces February 15, and their troops set fire to every warehouse where cotton was stored. In the turmoil of those final hours, a large quantity of ammunition exploded at the main railroad; 150 persons were killed and many others were injured. The city was surrendered to Union forces on February 18, a day after the fall of the City of Columbia.[7]

Northern troops under General Godfrey Weitzel entered the city, captured 500 guns, 5,000 stands of arms, and 6,000 prisoners. Thirty locomotives and 300 railroad cars were abandoned by the enemy, and ships of the Confederate fleet, at anchor in the James River, were destroyed.[8]

Colonel Elmer Ellsworth of the Fire Zouaves. When passing through Alexandria, Virginia he saw the Confederate flag flying from the cupola of the Marshall House, rushed into the building, up to the roof and tore down the flag. As he was carrying it down the stairway he was shot in the chest by the proprietor of the hotel. Ellsworth died instantly and a corporal immediately shot the proprietor, killing him.

Typical river scene during Civil War. As the USS Clifton, attached to the mortar fleet moves into position to bombard Forts St. Philip and Jackson, a Confederate fire raft drifts toward another ship. Note hose stream directed from Clifton. (U. S. Navy Photo.)

Major Nichols included the following impression in his report of Sherman's march through Georgia and other states:

When the army commenced its southward march Atlanta was given to the flames. On November 15, a grand and awful spectacle is presented with the city now in flames. By order, the chief engineer has destroyed by powder and fire all the storehouses, depot buildings, and machine-shops. The heaven is one expanse of lurid fire; the air is filled with flying, burning cinders; buildings covering two hundred acres are in ruins or in flames; every instant there is the sharp detonation or the smothered booming sound of exploding shells and powder concealed in the buildings, and then the sparks and flame shoot away up into the black and red roof, scattering cinders far and wide. These are the machine-shops where have been forged and cast the rebel cannon, shot and shell that have carried death to many a brave defender of our nation's honor. These warehouses have been the receptacle of munitions of war, stored to be used for our destruction. The city, which, next to Richmond, has furnished more material for prosecuting the war than any other in the South, exists no more as a means for injury to be used by the enemies of the Union.

From Rome, in the extreme northwestern corner of Georgia, close by the border of Alabama, draw a straight line southeastward. After three hundred miles it will touch Savannah. Then draw another line north one hundred and fifty miles, and it will strike Columbia, the capital of South Carolina. Thence draw another line northeastward two hundred miles, and it will touch Goldsborough, in North Carolina; from thence another line, drawn northwestward a hundred miles, will touch Raleigh and Chapel Hill, where the march really closed. In all there was a march of seven hundred and fifty miles in a straight line — but something more than a thousand measured along the roads actually traveled. A straight line drawn from Rome, the beginning of the march, a little north of east to Raleigh, its close, would measure about four hundred miles.[9]

Another aspect of the destructiveness of the Civil War is contained in the historical report of the Richmond, Virginia Fire Department which included this summary:

The worst of all Richmond fires occurred when the Confederate government and military evacuated the city. Under command of Confederate General Richard S. Ewell, the troops set fire to three warehouses, destroying the supplies they were forced to leave and then

A fleet of fire ships is massed against Union vessels near the "Head of the Passes" at the mouth of the Mississippi River October 12, 1861. **(U. S. Navy Photo from Harper's Weekly.)**

igniting Mayo's Bridge behind them. A high wind rose and spread the flames — the lawless element of the city surged through the streets, some no doubt setting fire to more buildings, and looted everywhere. The flames followed, and the great heart of Richmond, before the eyes of the citizens that were left here, passed in a night into utter ruin. From the northside of Main Street to the river, between Eighth and Fifteenth and from Twentieth to Twenty-third, only the Post Office Building (since torn down for the present one) and the Bank Building, yet standing next to Eleventh and Main Streets, were left.

So it was that the Federals, whom Richmond had fought to keep out of the city, at last arrived, not as destroyers, but as preservers of the city's peace and property. The dynamiting of buildings at strategic points had the effect of checking the fire, and many of the marauders were rounded up and imprisoned — at last the harassed city found peace in her desolation.[10]

One of the most spectacular and most critical battles of the Civil War was the attempt of the naval fleet under Admiral David Farragut to break through Confederate control of the Mississippi River and capture New Orleans. From the perspective of military and naval strategy and potential psychological rewards of victory, this was a vital mission for the Union forces. The South had three major forts protecting the opening of the river to the city of New Orleans since the river was a vital means of supply to inland troops. The Confederacy had made every possible arrangement to protect it from the stronger naval forces of the North. But on the evening of April 23, 1862, Lieutenant Caldwell of Farragut's fleet made a nighttime reconnaissance in a ten-oared boat to investigate the massive chains and other obstructions that were blockading the river. Apparently this reconnaissance was expected because the

For more than two centuries volunteer fire companies in America have conducted a great variety of fund raising activities to purchase appartus and equipment. Typical of the imagination and persuasion in these public occasions was the event conducted by the volunteer fire department in Louisville, Kentucky, in 1807. This was the occasion for display of the first elephant imported into America by a firm named Arnold and Company. Citizens who came to see the animal were assessed ten dollars for each showing. The money collected was used to purchase ladders for fire service!

Wilmington, DE

In Wilmington, Delaware, on December 22, 1775 a number of citizens organized the Friendship Fire Company No. 1 to protect the Borough of Wilmington from fire. Each member of this company pledged to furnish one large wicker basket and two small leather buckets for fire fighting. Each member also wore a red leather chest protector and a high red hat. At the end of December the Borough Council of Wilmington purchased two Newsham engines from a French man of war lying at anchor at the Port of Wilmington.

On September 11, 1860, just before the Civil War, the Exempt Association in Newark, New Jersey received this first-class Amoskeag steamer, designated "Minnehaha." It was assigned to the quarters of Engine 6. By that time, Newark had thirteen fire companies in service.

Confederates lighted immense piles of wood along the shores and turned loose a group of burning fire rafts. The latter were intended to block or burn the expected invading ships. Despite this action, the reconnaissance crew was able to cut the barrier and open the way for the larger ships.

This invasion produced one of Farragut's famous comments, made under the stress of the initial attack. The mouth of the Mississippi River and some of the bay had been seeded with floating mines by the Confederate defenders. These devices were rather crudely built but they could explode on contact with a ship and cause severe damage. In naval terminology of that era, these mines were called "torpedoes" and their presence was determined by Union reconnaissance just before the invasion began. An aide reported this fact to Admiral Farragut and received the famous reply, "Damn the torpedoes! Full speed ahead!"[11]

Just before two o'clock in the morning of April 24, Farragut's flag ship, the *Hartford*, signaled to the rest of the fleet to begin assault of the river. The first group of eight vessels under the command of a Captain Bailey passed through some of the obstructions and headed for Fort St. Philip but very quickly they were subjected to tremendous fire from the shore battery. In addition they had to meet ships of the Confederate fleet including gunboats and rams. Of primary concern were the Confederate fireboats and fire rafts, already ignited and floating down the river. These were a major harassment to the wooden vessels of the Union fleet.

Many Civil War historical studies report this entire battle as one of the most violent and colorful incidents of the war. Some of Farragut's ships had mortars capable of hurling large caliber shells for 3,680 yards. The Confederate shore batteries also had large caliber guns and mortars from which huge shells rose in fiery trails, directed against the incoming fleet. The fire rafts on the river were pushed by small unarmored tugs with heroic crews who tried to ignite them at the most advantageous moment. One of these rafts pushed up against the side of the flagship *Hartford* and flames began to involve the woodened sides and the sails. Admiral Farragut was quoted as saying "Don't flinch from that fire, boys! There's a hotter fire than that for those who don't do their duty." But a short time later as the fire became more intense he shouted, "My God! Is it to end this way?"[12]

A moment in one of the tremendous battles of the Mississippi as a Confederate battery fires on slow moving Union ships below. The original color lithograph of this scene is at the Library of Congress and bears the title "Battle of Port Hudson" by Davidson. (U. S. Navy Photo.)

From this time on, the cannonade became stronger, sustained at a tremendous rate. The shore batteries and the forts soon were knocked out and by daylight every Union ship except three, made its way through the barricades and north of the position of the forts. In effect, the Mississippi River was now open to Union control and on the 25th of April the City of New Orleans was captured.

The Battle of Mobile Bay was similar to the New Orleans engagement, to the extent that it involved moving the Union fleet under the tremendous barrage of shore batteries, with the additional complications of fire rafts and metal clad ram boats. Farragut had the objective of silencing and capturing three major forts that protected this cotton heartland of the South: Forts Fisher, Powell, and Morgan. Using most of the vessels from the New Orleans engagement, he began his assault on Fort Powell August 5, 1864, and very quickly destroyed that emplacement. Fort Morgan was more difficult because Farragut's ships were hampered by the movement of fire rafts and because of the incredible amount of mortar and rocket fire exchanged between ship and shore. This has been described as the heaviest volume of artillery fire of any war in history up to that time. Fort Morgan was captured August 23, 1864, after a battle that lasted for twelve hours, with an estimated 3,000 missiles hurled from guns of both sides. Considering the slow methods of handling, loading and firing shot and shell in that war, the slow movement of ships, and the continual dangers of the long barrage, together with the flames and explosions of the fire rafts, it is no wonder that this battle achieved notable rank in the annals of history. Fort Fisher was not captured until January 15, 1865.[13]

Fires in Washington and New York

While this savage war was underway, in many other parts of the new nation, municipal growth and expansion of businesses and industries continued at an increasing rate. But in cities and towns severe fires continued to occur and there were accompanying developments among the larger fire departments.

In 1861, the City of Washington (later to become the District of Columbia) introduced the first bill for equipping the paid fire department as a substitute for the volunteer system then in operation. However, this bill was defeated because President Buchanan, his cabinet and other elected representatives were engrossed in the political problems that immediately preceded the war. On the morning of May 9, 1861, a severe fire occurred in a liquor store next to the old Willard Hotel in Washington. This brought assistance from 100 Ellsworth Fire Zouaves from New York City who were visiting the city prior to military service. This highly trained group performed some excellent fire duty as well as vigorous acrobatics to the amazement of the watching crowd. Later, they marched to combat as Union troops, but they suffered devastating slaughter and their beloved leader, Colonel Elmer Ephraim Ellsworth was killed by a southern citizen when Ellsworth tore down a Confederate flag.[14]

On December 30, 1862, the renowned Ford's theater was discovered on fire. The blaze originated in the cellar and in three quarters of an hour nothing was left but blackened walls, a situation that simply was beyond fireground control.

On February 10, 1864, a fire destroyed the private stable of President Abraham Lincoln, killing six horses. Like many other fires of the time, this proved to be the work of an arsonist.

On June 17, 1864, a disastrous loss-of-life fire occurred at the U. S. Government Arsenal where 108 girls were in the main laboratory making cartridges for small firearms. Somehow a quantity of fireworks placed outside the building became ignited and a piece of fuse blew into one of the rooms where twenty-nine young women were working. The cartridges ignited and a violent explosion occurred. In the ensuing moment of fire and panic twenty of these girls were killed and a number of others were badly burned. The disaster had a powerful effect on the people of the city.[15]

On March 31, 1865, a steam transport ship, the *General Lyon*, traveling from Wilmington to Fortress Monroe with between 500 and 600 passengers, caught fire off the coast of Cape Hatteras. Less than forty of those on board were saved. Most of the passengers were soldiers or refugees from the southern areas of war.[16]

The Washington Fire Department was authorized as a fully paid organization May 19, 1864, but on July 1 that same year it was expanded as a part-paid fire department with three engine companies and one truck company. The engines had three

The famous New York Zouaves under the command of Colonel Elmer E. Ellsworth achieved some spectacular fire fighting at the Willard Hotel in Washington, DC. These men were in excellent condition and were likely to perform unusual acrobatics as shown in this illustration.

paid men and six call men; while the truck company had two paid men and seven call men.

Another significant development occurred October 12, 1864, when the first fire alarm telegraph was installed in the city. Known as the "Crystal System" it included about twenty-five street boxes and remained in use until September 30, 1875, when a larger system was installed in the city.[17]

65

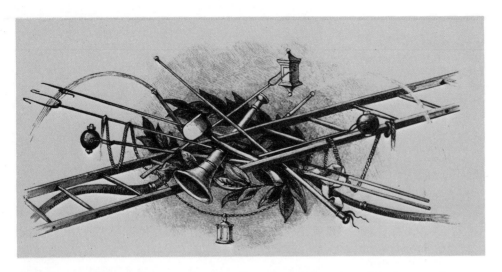

Equipment of the typical volunteer fire department in mid-nineteenth century including ladder, nozzles, hose, hooks, lamps, trumpet, axe and other necessities.

The Draft Riots[18]

Within the past two hundred years many cities, towns and rural areas have experienced the violence of angry mobs acting in a frenzy, but it is doubtful if any one of these incidents could match the viciousness and slaughter resulting from the draft riots of 1865 in New York City. This shocking example of mob insanity occurred during three days beginning Monday, July 13, and about 1,200 people were murdered or otherwise killed in the wanton fire destruction.

At that time lots were being drawn for conscription of civilians in military service in accordance with the Draft Act passed by Congress in the preceding March. New York City was designated to produce 18,523 additional soldiers through this means. Many citizens were extremely angry because people of wealth could avoid the draft by hiring substitutes to serve in their stead for the sum of $300, but those who did not have sufficient money were subject to induction.[19]

The resentful mob action began in the taverns and grog shops of waterfront neighborhoods, and was aggravated and otherwise helped along by southern sympathizers. The mob started up Third Avenue, beginning first with raids on stores and shops. At Forty-sixth Street and Lexington Avenue, Police Superintendent Kennedy, on an inspection tour, confronted the mob, was savagely beaten and left for dead, but he was only unconscious. Next the mob surrounded the Provost's office, where sixty police were on guard duty. They dispersed the officers, surged into the building, beat the Deputy Provost Marshal until he was unconscious then set the place on fire.

The temper of the mob was increasing. Dozens of them began setting fires in buildings on both sides of the street and within minutes the entire block between 45th and 46th Street was burning. When the fire alarm sounded and apparatus and fire fighters began to respond, the rioters grabbed axes from the hook and ladders and chopped hose and otherwise began to abuse the firemen. About that time, Chief John Decker, then in command of the department, arrived at the scene. A large, aggressive man, and a powerful speaker, he was able momentarily to shout down the mob and muster his fire fighters to the task of extinguishing the fire. However, some of the crowd had already started other fires two blocks away and, when Chief Decker and his men moved to this scene, other fires were started throughout the neighborhood.

The mob was completely out of control by this time, burning buildings, looting stores, fighting soldiers and civilians and beating policemen. The fire fighters kept racing back and forth in their apparatus trying to put out the fires as rapidly as possible. Finally night came and much of the violence subsided, but the mayor of the city was telegraphing desperately for help from military troops.

On the following morning, July 14, the mob again started its destruction, this time concentrating on colored people. These poor unfortunate citizens were routed out of their homes, their houses were set on fire, and then their bodies were thrown back into the blaze or hanged from the nearest lamp posts. A little later, in an orgy of violence, rioters charged toward the orphan asylum for colored children on 5th Avenue between 43rd and 44th Street. When they found the doors locked, they set the place on fire and Chief Decker raced to the building with some of his men. With a roar of rage, he charged the crowd, swinging a large nozzle to gain safety for his men for a few minutes and finally was knocked down by the angry mob, but meantime the children had been evacuated safely from the rear of the building. After this incident, the unruly crowd burned down the building of the New York *Tribune* and set fires all over Harlem. Authorities lost count of the number of murders and arson and the rioting continued until nighttime. Next day, a Colonel O'Brien of the 11th Regiment, was captured by the mob, beaten horribly and dragged through the streets for hours afterwards until he was dead.

Similar actions were going on in Brooklyn where the mob set grain elevators on fire and attacked fire fighters who responded. Finally, on the evening of July 15, military troops arrived, but order was not completely restored until the following day when another regiment marched into the city.

Without question, this three-day orgy of destruction and murder was the worst combination of mob action ever to be recorded in U. S. history.[20]

On July 18, 1865 the renowned museum of P. T. Barnum at Broadway and Ann Street was destroyed by fire. Members of the new paid fire department were able to confine the fire to the block, but the museum was completely destroyed in the presence of a huge throng of cynical observers. Shortly after, it was determined that this fire was a result of arson. Robert C. Kennedy, a spy for the Confederacy, was convicted in November of that year, of trying to burn New York City. He confessed to setting the Barnum Museum fire and five other fires which destroyed hotels. He was executed by hanging on the following March 25th at Fort Lafayette.

In the midst of the turmoil of the Civil War three new states were admitted to the Union: Kansas in 1861, West Virginia in 1863 and Nevada in 1864.

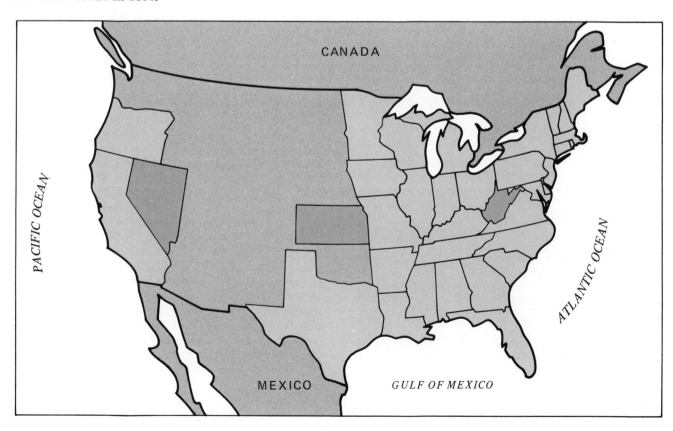

Alarms of Fire

Over the years different means have been used to notify the local fire department and the public that a fire was burning. Shouts, gunshots, the beating of drums, the ringing of bells, rattles, lights, whistle blasts, sirens and other signals were sounded or flashed in cities and towns throughout America.

Boston, Massachusetts, was the first to employ two bellmen in 1659 who were assigned to walk through and about the town from midnight to 5 a.m. The legislation enabling this action included this statement: "if he sees an extraordinary light or fire in any house or vessel, to inspect it, and if necessary, to give an alarm." The alarm usually consisted of much shouting and sounding of the wooden rattles carried by these bellmen.

New York, Philadelphia and other large communities soon began similar tours of nightwatch.

In Providence, Rhode Island, at the end of the 18th century, there were no street lamps in the highways, and, when a fire occurred, the citizens were supposed to place lights in their front window and every bell in the town was rung violently.[1]

Attendant striking bell station alarm after receiving telegraph signal or some other indication of fire location.

In Charleston, South Carolina, before the Revolutionary War, night watchmen manned the steeple of St. Michael's Church, looking for fires. When one occurred these guards or sentinels sprang to their rattles and yelled "Fire" to fire masters and engine companies. Men on horses then raced through the streets sounding rattles and yelling. St. Michael's bell rang for fifteen minutes as did chimes and bells in factories (daytime). "Everybody poured into the streets to increase the clamor" stated a historical report of the fire department.[2]

In Cincinnati, in 1802, fire alarms were sounded by the blowing of foxhorns and all physically able male members of each family were required to respond. In 1809, a big drum, five feet high and sixteen feet in circumference, was mounted on the roof of a one-story carpentry shop and the beating of this drum was the signal of a fire alarm or the toll of some other disaster. A ladder was kept in the rear of the building for persons to climb and sound the alarm. In 1824, this drum was replaced by several bells, the principal one being in the Presbyterian Church. In 1845, citizens erected a new watch tower on the roof of the Mechanics Institute and five years later the city authorized construction of an improved watch tower on the same building. In 1853, a fire alarm bell was added and in 1860 Cincinnati installed its first fire alarm telegraph system.[3]

Sioux Falls, South Dakota, in 1884, bought a large bell for a fire alarm. This weighed more than a ton, 2,600 pounds actually, and was mounted in the fire tower of the city auditorium. Later it was moved to the tower of the central fire station.

Boston, Massachusetts, was reluctant to let tradition end, and its fire alarm practices continued for a long time. In the 19th century and well into the 20th century, bells in church towers were used for alarms of fire. In due course, members of the fire department were given duty of attending to the church clocks and bells, many of which were later arranged to be rung by the fire alarm system. This chore continued into the middle of the 20th century; in fact, on October 26, 1958, as Daylight Savings Time was coming to an end, a Boston Fire Department order contained this message: "All tower clocks without striking mechanism will be set back at six-thirty a.m. All tower clocks with striking mechanism will be stopped at six-thirty a.m. for one hour, then started. Only men who are regularly assigned to wind clocks should be detailed to change tower clocks."[4]

Many communities relied upon mill or plant whistles to give the location of a fire by a series of blasts and this practice is still in use in some areas. During the "Cold War" of the 1950s, Civil Defense funds were used to install alarms and sirens for this purpose in thousands of communities.

In the City of Los Angeles, just about the start of the 20th century, the customary practice of signaling fire emergencies was to shoot a gun, then to shout "Fire!", which would usually be followed by someone running to the nearest engine house and ringing a bell.[5]

The Baltimore, Maryland Fire Department up until January, 1893 rang bells in fire house towers in the City Hall as a means of summoning call members of the department. On January 1 that year, the department became fully paid and the bells were no longer needed.[6]

Fargo, North Dakota, used the whistle at the water works to signal fire. Usually the superintendent of the water works was in the vicinity to pull the whistle cord.[7]

In 1863, the city council of Indianapolis, Indiana, ordered the fire tower placed on a three-story building in the center of town. This was manned by a watchman whose duty it was to look over the city and watch for fires. The tower was in a fairly elevated position since none of the surrounding buildings were more than three stories high. A bell was placed in open frame-work immediately in the rear of this block and was rung by means of a device connecting it to the tower.

At that time Indianapolis was divided into nine loosely settled wards and for five years the location of a fire was designated by the towerman striking the number of the ward. This gave a general designation of the fire location but very frequently responding fire companies would have to search through the district to find the fire building.

During the Civil War the watch tower was used for another purpose, as explained in this extract:

"When Confederate prisoners were confined at Camp Morton (afterwards the State Fair grounds, and now Morton Place), the man on the tower had special orders to keep his field-glass leveled on this prison camp, two miles and more to the north. The city was kept in a constant state of apprehension by frequent rumors of conspiracies for the release of these prisoners, and of threatened outbreaks, on which occasions it was popularly held that it would be the purpose of the prisoners to set fire to the city, which they would then pillage, and in the meantime murder all the inhabitants. At the first appearance of danger it was the duty of the watchman on the tower to sound the signal, whereupon the citizen soldiery were to assemble at the armory and get their guns and munitions of war. It was struck more than once in the feverish days in 1863, when Confederate raider John Morgan invaded the State. He was supposed to be advancing on Indianapolis, but as the telegraph lines were cut, there was great uncertainty as to his

The old castellated watch tower in Cincinnati with the huge bell at right.

In 1938, Kingston, Ontario celebrated its Centennial, decorating this typical end-of-the-century fire station, topped by a watch tower.

movements, and the rumor that he was within a few miles was enough to almost incite a panic. As to the prisoners, the firemen had orders to keep a close watch upon the hose and engines; in case of an outbreak at Camp Morton there would, in all likelihood, be an attempt by the Confederates or their sympathizers, after they had fired the city, to cripple the department by cutting the hose or injuring the engines. For a year or more every member of the department carried a revolver for use should any need therefor arise."[8]

Many small towns used a large iron hoop that was struck by a sledge hammer to sound the alarm of fire. Today, quite a few fire companies still maintain these contrivances in prominent position in front of the fire station.

In Phoenix, Arizona alarms of fire were sounded by three pistol shots in rapid succession. Later, the steam whistle of a local flour mill was used, then, in 1884 the bell in the tower of the county courthouse was relied upon for the sounding of alarms.[9]

Newark, New Jersey perhaps had claim to the largest alarm bell. From 1864 to 1898 this 11,000-pound giant was in the tower of the Hose Depot Building, which later became Fire Department Headquarters. At first it was manned by a watchman, later by fire fighters, It rang box numbers to signal locations of alarms; a noon-time signal so people could set their watches and clocks; and then a 9 p.m. curfew when fire horses were bedded for the night and children were required to be home.

The bell had to be replaced several times, but gradually the metal tower began to deteriorate and in May, 1898 the bell and the tower were sold. By then, the city had an extensive telegraph fire alarm system and citizens could also call in alarms by telephoning Number 25.[10]

New York City had a fire bell installed on top of its city hall in 1830; later divided the city into pie-shaped districts radiating from this tower. The first coded fire alarm signals were reported to have been struck from this location, according to the district locations.[11]

Post Office bell tower in New York City. In 1847 Chief Engineer Anderson urged installation of a telegraph system to be connected to these bells, to cut down on the frequency of false alarms.

Central fire station in Fargo, North Dakota before it burned in conflagration of 1893. Tall structure at left is bell tower for signaling fire alarm.

The Self-Propellers[1]

Engine No. 35 of Boston, Massachusetts, an Amoskeag self-propelled steam fire engine, on a trip to a fire.

The first steam fire engine in the United States was designed and constructed at the instigation of insurance companies in 1840 by P. R. Hodge, C. E., an ingenious mechanical engineer of New York City. This was an odd-appearing self-propelled affair about 14 feet long, shaped like a locomotive, with huge driving wheels at the rear. It weighed between seven and eight tons. The engines were horizontal and had cylinders attached to the smokebox at the front end of the boiler. The pistons of the steam cylinders and water pumps were on the same rods and the connecting rods were attached to cranks on the wrought iron rear wheels. When the engine arrived at the scene of the fire, the rear end was raised clear of the ground by means of a jack-screw and the wheels used for balance.

Although this engine apparently passed tests, it was not a complete success. It was cumbersome, slow to get into action and frequently (so the chief reported to the Common Council in 1841) some part of the machinery broke down after being set in motion. Moreover, its presence was bitterly resented by the volunteer firemen who pinned their faith to manpower at the brakes of hand engines. So it was put aside and not until 12 years later were property owners and insurance underwriters convinced that a more efficient means of extinguishing fires than manual engines was needed.

The next attempt at building a steam fire engine was in 1852 when Moses Latta of Cincinnati designed and built the first *successful* steam fire engine, the "Joe Ross." This also was a ponderous self-propeller with two large rear wheels for propulsion and a smaller one in front which served as a steering wheel.

Like the Hodge engine, the "Joe Ross" was odd looking. The boiler was square and upright and known as a "gunpowder" because steam was generated by flashing cold water into a hot boiler. It had two horizontal cylinders; one on each side. The pumps, as on the Hodge engine, were in front of the cylinders and driven direct. The engines (cylinders) were coupled

P. R. Hodge built this steam fire engine in 1840 to launch the era of the "self-propelled." Note jack device between rear wheels at right. During fire fighting, this was used to raise the rear end clear of the ground. Generally, though, this engine was unsatisfactory and it was another twelve years before the "Joe Ross," the first successful steam fire engine saw action.

71

Similar to the renowned "Joe Ross" of Cincinnati, this "Citizens' Gift" was built by the Latta firm in 1854 through contributions of the people of that city. Note square boiler and suction hose.

with the driving wheels at the rear by connecting rods. Altogether, the "Joe Ross" weighed more than ten tons and required pulling by four horses, (ridden artillery fashion), and propulsion assistance by the engine.

At the acceptance test, the engine was ready to work in seven minutes from the time the fire was lighted under the boiler until water was thrown through a 3-in. hose with a 1½-in. nozzle to a distance of 225 feet. Later, six good streams were obtained at one time using small calibre nozzles.

This self-propeller was accepted on January 1, 1853, and remained in service until December 6, 1855, when, while being tested, its boiler exploded killing the engineer and injuring several others. The explosion was due to the firebox not being properly stayed. A steam pressure of 180 psi was being carried at the time.

In 1854 another self-propeller, similar in construction but weighing only about half as much as the "Ross," was built from voluntary contributions of the Cincinnati citizens. It was appropriately named "Citizens Gift," had double 11-in. cylinders and 6⅝-in pumps and threw a stream through 2½-in. hose with a large nozzle a horizontal distance of 297 feet. Its record for raising steam to a working pressure was 3 minutes 10 seconds.

The firm of A. B. and E. Latta did quite an extensive business but how many of their engines were self-propellers is not known. In 1863 the business was sold to Lane & Bodley who built seven or eight machines among them at least one self-propeller of an improved Latta pattern for Engine No. 10 of St. Louis, Missouri.

Another early builder was the New York City firm of Lee and Larned which built three self-propelled engines in 1859. One went to Exempt Engine Co. of New York City, one to Southwark Engine Co. of Philadelphia and the third to Pittsburgh, Pennsylvania.

These were known as mongrel engines because they had rotary pumps and piston cylinders. They resembled street locomotives and were capable of propelling themselves on any ordinary street or road. The power to drive the machines on the highway was transmitted from the engine (steam cylinders) to the rear wheels by parallel rods which, on reaching the fire, could be disconnected instantly and the power transmitted directly to the pump, also at the rear.

The boilers were upright, located about midway between the front and the rear. The frame, made of boiler and angle iron, was suspended on four lengthwise springs and one crosswise spring under the rear axle. The pump could discharge six gallons per revolution and at ordinary working speed would discharge from 600 to 700 gpm.

The engines were guided by the running gear of F. R. Fisher whose steam carriages were attracting a great deal of attention at the time. The two front springs, placed one above the other in the line of the center of the carriage, took hold of a vertical spindle connected with the forward axle by a kind of universal joint. At the top of this joint was a horizontal crank, which, when turned by a worm and screw, controlled the direction of the axle, thus permitting the engine to be steered easily and precisely.

In 1859 the firm of Lee and Larned in New York City built three self-propelled engines. This one went to the Exempt Engine Company.

These engines did not meet with the general approval because they were heavy and cumbersome. After the first three were built, the company built lighter engines to be drawn by men or horses.

The only company which built successful self-propelled steam fire engines was the Amoskeag Manufacturing Company of Manchester, New Hampshire, and its successors the Manchester Locomotive Works and the International Power Company. The first Amoskeag self-propelled engine was constructed in 1867; the last in 1908. In the years between, the firm built twenty-two self-propelled engines. All gave satisfactory service.

The original idea for a self-propeller started from a little steam-propelled wagon built by two foremen at the Amoskeag plant, James Bacheller and W. H. Rittener of Manchester, in 1862. This wagon attracted the attention of Freeman Higgins, superintendent of the machine shop, and Daniel Morse, superintendent of the brass foundry, and as a result of a discussion among the four of them, two experimental steamers were built. On its first trial, one of these steamers ran from the Amoskeag plant up the Stark Street hill to the fair grounds at a speed of about eight miles per hour.

One of these Amoskeags had its rear wheels solid on the axle; that is, it had a round axle which turned in two boxes on each side of the frame. It was of the straight frame type (the front wheels did not cut under) and had a chain drive extending from the pump shaft to each rear wheel; an arrangement which enabled the engine to propel itself fairly well although some difficulty was experienced in turning corners. The other engine was of similar construction but only on one rear wheel, solid on one axle, free on the other. It, too, was difficult to handle in turning corners.

One, or maybe both, of these engines was demonstrated in Boston and other cities but there was no particular demand for self-propelled engines and they remained in the Amoskeag shops for some time. In 1872, the horses in the eastern part of the country were attacked by an epizootic disease and wagons frequently had to be hauled to fires by men and boys. As a result of this situation, New York City purchased one of the Amoskeags in October, 1872, where it was assigned to Engine Co. 20. The other was sent to Boston (by train) to help fight the conflagration of November 9, 1872, where it rendered effective service. Following the fire it was purchased by the city and placed in service with Engine Co. 21 in the Dorchester district. It was rebuilt as horse-drawn apparatus in 1874 and remained in active service until 1894.

While on a business trip to New York in 1873, Superintendent N. S. Bean of the Amoskeag Company met a gentleman who suggested that self-propelling engines would be greatly improved by incorporating in their driving mechanics a "compound" gear such as was in general use on cotton speeders in cotton mills. The gentleman claimed that by use of this gear, power could be applied to both driving wheels in such a way that when turning corners, the wheels would be driven automatically without appreciable loss of power.

Maneuverability of Amoskeag steamers was increased with the introduction of compound gears in the rear axle. This unit was the first to feature this improvement and went to Lafayette Engine Company No. 1, Detroit, Michigan.

This trolley car was used to transport steam fire engines on out-of-town mutual aid calls. This is an 1871 Amoskeag from Engine Company No. 1 in Springfield, Massachusetts.

Years of service were given to Boston, Massachusetts by this Amoskeag Engine 38 shown here in 1897 and finally retired in the 1920s.

Boston's Engine 38 in a pumping exercise. Note shovel and poker leaning on rear wheels.

Bean, having other matters on his mind, did not give any particular thought to the suggestion and did not even learn the gentleman's name, but after his return to Manchester he came to the conclusion that the idea was worth looking into. As a result, one of the gears was manufactured for a steam fire engine.

At the time, a self-propelled engine was under construction for Lafayette Engine Co. 1 of Detroit, Michigan, which, except for a crane neck frame in place of the straight frame, was of the same general design as the engines then in service in Boston and New York. It was decided to make the experimental installation on this engine. In due time, the engine was completed and tested. Its trials proved that the suggestion of the gentleman in New York was sound in every particular and that the self-propeller feature had been greatly improved. All engines subsequently made were equipped with this device which was patented by Mr. Bean in 1875.*

The Amoskeag catalog for 1874 commented on the driving mechanism of these engines as follows:

The propelling and steering apparatus are very simple in this construction, and are so arranged as not to interfere in the least with the use of ordinary drawing for either horses or men, if it is desirable to use them. The propelling is done by the same wagons that are used for the pumps and these are made reversible, so that the machine can be propelled backwards and forwards, as desired.

The engine is propelled by a connection between the crankshaft and the rear wheels of the engine, made through an endless chain working over sprocket wheels, with a set of compound gears upon the rear axle so arranged that in turning the engine the two rear wheels are driven at varying speeds.

The steering apparatus is arranged so that while the driver is enabled to change the inclination of the wheels at will, the jar occasioned by locomotion is inoperative to change the position of the steering wheels; hence, the constant exertion of the drivers is not required to keep the machine in line even when moving at the rate of 16 miles the hour.

Between 1873 and 1889, thirteen Amoskeag self-propelled engines were built, all approximately like the Detroit engine. In 1889, a double extra first size self-propeller was built for Hartford, Connecticut, which formed the model for all later engines of this type.

The 1889 Hartford and subsequent engines were constructed with rod and cross-head connections between the steam cylinders and pumps which made it possible to change or reverse the motion of the engine. The road driving power was applied from one end of the main crankshaft, through a "compounder" or equalizer, to two endless chains running over sprocket wheels on each of the rear wheels of the apparatus. This permitted the rear wheels to be driven at varying speeds, when turning corners, without loss of power. When it was desired to work the pumps, the engine was disconnected from the road driving gear by the simple expedient of removing a key.

---

*It is interesting to note that the differential gear with which all modern automobiles are equipped is essentially the same device, although differing in design and detail, as that applied to the self-propelled steamer fire engine Lafayette No. 1 of Detroit, Michigan, in 1873.

Control equipment was located at the engineer's post at the rear of the boiler and was similar to that of a locomotive with throttle, reversing gear, and other controls close at hand. Steering was effected by a large hand wheel at the front of the machine and by action of a pair of bevel gears and a worm gear connected to the front axle. It required very little effort on the part of the steersman to keep the engine in line or to change the inclination of the wheels even when traveling at full speed. On a fair road the vehicle could make over ten miles per hour. Spike-tires were provided for winter service.

The Hartford engine, built in 1889, was the city's third self-propeller and was considered the largest and most powerful engine in the world.

It was assigned to Engine Co. 3 located in the heart of the downtown section and was known far and near as "Jumbo." Its height to the top of the smokestack was 9 ft. 9 in.; length overall was 15 ft. The boiler was 40 in. in diameter and contained 301 copper tubes. The pumps were double; 5½-in. in diameter with 8-in. stroke. The steam cylinders were 9½-in. with 8-in. stroke. Its pumping capacity was 1300 gpm.

This engine was in active service as a self-propeller until 1915 when it was converted to a tractor-drawn machine by the removal of its self-propelling mechanism and the addition of an A & B gasoline-electric tractor. It was placed in reserve in the late '20's and is now in an automobile museum in Arkansas. Another Hartford self-propeller, Engine No. 4 built in 1901, is still in existence at Wilson, Connecticut, in its original condition.

Most of the Amoskeag self-propellers lived long and useful lives and some of them remained in active service after motor apparatus came into general use. New York City purchased its first one in 1872 and four more in 1874. They remained in service with Engine Cos. 8, 11, 24, 31 and 32 until 1884 when two of them were converted to horse-drawn and the others placed in reserve. Two others built in 1874 for Brooklyn, New York, and Fond du Lac, Wisconsin, were converted to horse-drawn engines in 1896 and 1897 respectively.

The two last Hartford self-propellers, built in 1889 and 1901 respectively, were not retired until the late '20s; the two Boston engines (Engine 35 and 38) were removed from active service in the '20s but Engine 38 was still in the fire station on Congress Street in the late '30s and the writer saw the Portland, Maine, engine, built in 1903, in the Central Fire Station in the early '40s. The last Amoskeag self-propeller built went to Vancouver, B.C., in 1906.

Much of the service performed by the later self-propellers was at large fires where their pumping capacity of 1300 gpm was needed. Their use on ordinary alarms had one serious drawback; sparks from their smokestacks as they roared down the street frequently ignited awnings and other combustible trim on stores and other buildings along the way. It was stated by Boston fire officials that on a summer day it was necessary to send a chemical company along behind Engine 35 from the Mason Street fire station as it traveled along Tremont Street en route to a fire. Hartford,

This 1901 Amoskeag self-propelled steamer was used by Engine Company No. 4 in Hartford, Connecticut. As of 1974 it was still on display in a Wilson, Connecticut fire station.

where two of these engines responded to most alarms in the business district, reported a similar experience. In Portland, Maine, the company to which the self-propeller was assigned also had a horse-drawn steamer. The former was used at big fires only; the latter for all first alarm response.

In 1903 Engine 5 of the Portland, Maine Fire Department had this Amoskeag steamer, No. 772 in the group of more than a thousand steamers produced by that company.

The Baltimore Maryland Fire Department in 1886 posed in front of city hall with this 1849 Agnew hand pumper.

Hand Engines

These pages show a variety of hand engines used by fire departments during the eighteenth century. They include the simple, bucket-filled type in use at the start of the century, and the heavy-duty, ornately decorated engines so highly prized by the fire fighters.

Note the decorations on Old "Black Joke," No. 33, another handtub made by James Smith and used and serviced in 1824. (Smithsonian Institution.)

Bucket-filled hand engine used in Beverly, Massachusetts in early 1800. It was very similar to the Newsham.

Hand tub presented to city of Toronto, Ontario in 1837. It was used during rebellion of that year after local bridge had been set on fire. Note hand brakes, suction hose and wooden wheels. (Toronto Fire Department Photo.)

This Hunneman hand engine in The Dalles, Oregon was purchased by Dallas City in 1862. Originally it was built in 1847 for the Medford, Massachusetts Fire Department and was designated as General Jackson No. 2.

Here is a handsome model of a hand pumper made by L. Button Company. It was in use about 1855. Designated the John B. Chace 4, its large suction hose is on top between the hand brakes bearing the slogan "Our Duty is Our Delight". (Smithsonian Institution.)

New Haven used this hand-drawn pumper from 1835 to 1860. The fire department began as a volunteer group about 1822, comprised of businessmen and other prominent community leaders. It became a paid fire department in 1862.

Here is the rotary hand pumper used by the fire company in Cortland, New York in 1803. (Smithsonian Institution.)

An 1857 hand pumper of the Louisville, Kentucky Fire Department. (Louisville Fire Department Photo.)

This is Americus No. 6, a hand pumper made by James Smith about 1851. (Smithsonian Institution.)

Lafayette Engine No. 4 in Newark, New Jersey about 1853. This was a Philadelphia style second class engine built by John Agnew and decorated with portrait of Benjamin Franklin. Ironically, one of the pumper's builders, foreman Jacob Allen was killed by a falling chimney, the first fatality in the history of the Newark Fire Department.

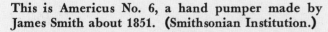

The Crosby Opera House burns during the great fire that devastated the city of Chicago, Illinois in 1871.

1866-1900

Chapter IV

. . . Later it was determined that there was ample time for everyone to escape from the building, but the frightened people packed into the stairway, screaming and fighting in a struggling mass . . . .[1]

Immediately after the Civil War there was a great surge in population movement as the nation gathered strength for redevelopment, resettlement, and industrial expansion. Many Americans were traveling to undeveloped lands in the central, southwestern and western parts of the country. Larger cities and towns in the north and south began rebuilding and expansion programs to recover from the economic restrictions of the four-year war.

Those next three decades, until the 20th century, brought major changes in fire protection throughout the land. In the preceding years, many thousands of buildings had been constructed with highly combustible exterior materials — wood shingles and clapboards, or other flammables. These were contributing factors in enormous conflagrations that swept municipalities in the post-war era, arousing national interest in the control of fire, and the gradual application of technical knowledge to solving the fire problems.

Between 1860 and 1900 urban areas in America grew twice as fast as the population. Chicago, a city of only 30,000 population in 1850, doubled its population every ten years until it included millions of citizens. New York grew to become the second largest city in the world. The same kind of expansion affected most of the cities and towns, even as the war was underway.[2]

The war also stimulated the country's ability to produce new materials and to utilize natural resources. Steel, petroleum, electricity, coal and iron, and the railroad became focal points for production and capital, but the concentration of buildings in the major industrial and mercantile areas created the conditions for future holocausts.

Fire departments were improving in capability — manufacturers were developing new steam-powered pumps and more functional apparatus and before long the first mechanically operated aerial ladder truck and other mobile apparatus became available. Means of communications were expanding with the innovation of telegraph and telephone systems which brought quick alert signals for practically every emergency. This era was to signal the end for some volunteer fire companies in the major cities, as the need became obvious for fulltime, paid personnel, manning fire stations around the clock.

By this time, the entire economy and productive capacity of America was benefitting from the technological upsurge in American industry which was tolling the end of past traditions and signaling a completely new era — the 20th century.

Major Fires

The Civil War had not quite ended when a destructive fire occurred April 1, 1865 in the Ordinance Office on Pennsylvania Avenue in Washington, DC. Starting in the second story of the building, it spread quickly to other structures nearby. Military personnel began carrying out furniture and papers but the wood-framed building burned rapidly. Directly across the street from this structure were the Government Work Shops, also frame buildings, and they were in great danger as wind carried sparks from the involved buildings. Howevery, military guards and fire companies from DC arrived promptly and managed to protect exposures. Washington had received its first fire "Crystal" telegraph fire alarm system October 12, 1864, and about twenty-five street fire alarm boxes were in use. The first alarm for this incident was sounded from Box 42, then telegraphed from Central Station to the large alarm bells in individual fire stations. Response was prompt and the endangered frame buildings of the Government Work Shops were protected by good fire fighting action.

The alarm system also proved important during the great Baltimore fire of July 25, 1873, when Washington sent two engine companies on a railroad train to that endangered city. Each company had its horses, hose carriages and 900 feet of hose. The steam fire engine from the Navy Yard was assigned to cover the station of Engine 3 on Capitol Hill during this emergency with the agreement that, "in case of fire at the Navy Yard it will be summoned at once by telegraph."[3]

In 1871, Chicago, Illinois, and the small lumber community of Peshtigo, Wisconsin, suffered the most dramatic fires of the century (see pages 83 and 87). Boston (see page 89) and Baltimore in 1872 had similar disasters.

Richmond, Virginia again had a severe fire in 1882, watched the destruction of the "Petersburgh Bridge" in 1883, had a severe fire in the State Penitentiary in 1888, and another blaze in the locomotive works in 1893.

The Chicago, Boston and Baltimore fires were responsible for more than seventy insurance companies going into bankruptcy.

New York City had some terrible loss-of-life fires in this period: the Brooklyn Theater fire December 5, 1876 took the lives of 295 people (see page 94); the Taylor Building fire August 22, 1891 took sixty-one lives; the Hotel Royal fire February 6, 1892 accounted for twenty-eight lives; the Windsor Hotel fire March 7, 1899 took forty-five lives.[4]

An 1885 fire fighter holding a trumpet. His uniform, typical of the times, includes a helmet, big-sleeved shirt, heavy belt and buckle, trousers and boots. (Smithsonian Institution Photo.)

Race with Flaming Death![5]

A railroad engineer's experience with a fire-involved cargo of flammable liquid.

Back in 1869 I was a hogger on a mixed run of the Buffalo, Corry & Pittsburgh between Mayville Summit and Brocton, New York. The two towns are only ten miles apart, but the track between them is so crooked that you have to travel fourteen miles to get there by rail, and the grade, over the whole distance, is over forty feet to the mile.

About nine o'clock on the night of August 17, 1869, we reached the Summit with a train of two passenger coaches carrying between fifty and sixty passengers, six wooden oil cars loaded to their brims, and a boxcar containing two valuable trotting horses and their two keepers bound for a race in Cleveland.

A few minutes later, I got a signal from the conductor and pulled out for Brocton. We were under good headway, when I glanced back and saw that one of the oil cars in the middle of the train was on fire. I reversed the engine and whistled for brakes. Both the conductor and the brakeman jumped off, uncoupled the coaches which were on the rear of the train, and set the brakes to hold them on the grade, bringing them to a stop.

Assuming that the oil cars had also been braked, I called to the brakeman on the boxcar just behind us to draw the coupling pin separating the car from the head oil tank, and backed up to take up the slack. What I had in mind was to run far enough away to save the boxcar and the locomotive.

When the pin was drawn, I ran the train down hill a piece to get out of danger. On looking back, I was horrified to see all six oil cars rolling after me with rapidly growing speed. Either the brakes had not been set or they had failed to hold. I didn't know which. But I did know that all I could do now was to make a run for it into Brocton, though at the speed we'd be obliged to take those sharp curves — twenty miles an hour was usual on them — the chances were we'd never get there.

When I saw those flaming cars — all six of them were on fire by now — racing toward me only a few rods away, I pulled the throttle wide open. But not soon enough. The oil cars caught me before I got underway, slamming with full force into the rear of the boxcar, smashing in one end of the car and knocking the horses and their keepers flat on the floor.

The heat was almost unendurable. No matter what I did, I couldn't put more than thirty feet between the pursuing cars and ourselves. By the light of the firebox as my fireman opened the door, I saw the face of one of the keepers in the boxcar. He had climbed up on the grate opening. He begged me, for God's sake, to give the engine more steam.

I was giving her all the steam she could carry. The grade itself was enough to carry us down at fifty miles an hour. Every time we hit one of those curves the old girl would run on just about one set of wheels. Why in the world she didn't topple over is something I'll never understand. Maybe the speed kept her on the tracks. Anyway, it was as if she realized it was a race between life and death, and she struggled desperately to save us.

The night was dark, and the railroad ran through woods, deep rock cuts, and along high embankments. We thundered along through them at lightning speed, with that hell on wheels only a few feet behind us. Those wooden tanks contained at least 50,000 gallons of oil, and with the rush of the wind against them they trailed a flying avalanche of fire three hundred feet along behind them. Flames shot nearly a hundred feet into the air, and the roar sounded something like a giant cataract. Now and then a tank would explode with the violence of a cannon, send a column of flames and pitchy black smoke mounting high above the cars. Burning oil scattered about in the woods, and small forest fires were springing up along the track behind us.

At the speed we were making, it wasn't long before the lights of Brocton came into sight down in the valley. I was relieved. For a moment I thought we might even make it. Then I remembered that Train No. 8, the Cincinnati express, on the Lake Shore, was due at the junction about the time we would be there.

Our only hope all along had been that the switchman at the junction would think far enough ahead to open the switch connecting us with the crossover track and let us run in on the Lake Shore track where the grade would be against us and enable us to get away from the oil cars.

But now the switch would, of course, be closed for the express. Our last hope was gone unless the express was late or unless someone had enough sense to flag it down.

Just then my fireman and I saw the express tearing along toward the junction. In those next moments I lived years, waiting to find out whether we would reach the junction and get the switch in time for it to be set back again for the express. If we didn't, we would crash, and scores of people, passengers and crews, would be killed or injured.

"Good God," I said to my fireman, "what are we going to do?"

The fireman — he was a brave little fellow — said all we could do was whistle for the switches.

That whistle was one prolonged yell of agony, a shriek that seemed to tell us our brave old engine knew our danger and had her fears.

Neither the fireman nor I spoke another word. Then, the engineer on the express train, seeing us tearing down the hillside with that long trail of fire behind us, realized that only one thing could save us. He whistled for brakes and brought his train to a stop not over ten feet away from the switch points. The switchman now understood our signal and opened the switch. We shot in on the Lake Shore track and whizzed by the depot like a rocket. The burning oil cars followed us in. But the race was over. Their momentum carried the oil cars after us up the track for nearly a mile before they came to a stop. For three hours more they continued to burn, sending a pillar of smoke and flame high into the night sky. And then there was nothing left of the six wooden cars but a heap of smoking ruins.

When we finally brought the locomotive to a stop, my fireman and I were both so weak that we couldn't get out of the cab. The two horsemen in the boxcar behind us were unconscious, and the horses were ruined.

But we were all alive.

And how long do you think it took us to make those sixteen miles from Summit to the spot two miles up the Lake Shore siding where we stopped? Twelve minutes — at a speed of eighty miles an hour! But those twelve minutes were the longest and most horrible I ever lived.

(Duff Brown died in 1884, leaving the above true story of his most exciting experience.)

Scenes from the Great Chicago Fire

WHERE THE FIRE BEGAN.

POST-OFFICE AND CUSTOM-HOUSE.

CHAMBER OF COMMERCE AND COURT-HOUSE.

LAND-OFFICE, ILLINOIS CENTRAL R.R.

CROSBY'S DISTILLERY.

REPUBLIC LIFE INSURANCE COMPANY.

MASONIC TEMPLE, DEARBORN STREET.

FIRST NATIONAL BANK.

Scene at the edge of the Chicago River when grain elevators, ships and other occupancies were involved by the sweeping fire.

The Great Chicago Fire[6]

Perhaps the most well-publicized fire in U. S. history was the one which destroyed the City of Chicago, Illinois, on October 8, 1871. Helped by a strong wind, this conflagration spread through about 2,150 acres of the city, destroying thousands of buildings and causing a loss of human life estimated at about 300. It was one of a series of conflagrations that were to devastate U.S. cities during the next forty years, but because Chicago was a new, important, rapidly growing midwestern center, this fire tragedy sent a shock wave of immediacy through the nation.

Weather conditions in the summer and fall of that year should have provided some foreboding of disaster. Chicago had only twenty-eight percent of its normal summer rainfall and, shortly before the conflagration occurred, conditions in surrounding states were highly conducive to wood fires. In the first week of October the Chicago Fire Department responded to twenty-seven fires. The worst of these occurred Saturday night, October 7, and it burned for fifteen hours, sweeping through twenty acres of the city before it was finally stopped by the underequipped fire department.[7]

It began in a planing mill on the west side of the city and consumed practically every building in a four-block area before it was brought under control Sunday morning. One person was killed and several others were injured and the fire department also lost hose and apparatus, including a steam fire engine and a hose cart. At that time Chicago had 185 fire fighters on the roster and half of them remained on duty all day Sunday as embers of this fire continued to smolder. They were not prepared physically or emotionally for the tremendous conflagration that was about to begin.

As we look back on that tragedy, we must consider the limitations of fire protection and fire control that became so important as the conflagration developed. The fire department had seventeen steam pumpers, fifty-four hose carts, four hook and ladder trucks, two hose elevators and forty-eight thousand feet of fire hose. The city, running in a longitudinal direction south from Lake Michigan, had three divisions, the South, the North, and the West. The South Division contained the principal business district and was bounded by the south branch of the Chicago River and Lake Michigan.

The North Division was bounded by the north branch of the river and the lake. The West Division included all of the city west of the river. Each division had residential property, mostly of wood frame construction, with wood siding and wood shingle roofs. There was hardly any zoning of properties (as we understand the purpose of zoning today), no regulated spacing between buildings, no automatic extinguishing systems, and little or no control of building construction. Another matter of importance was that the newspapers and other media of the time contained few facts or other information that might warn the public of impending disaster.

Many people have studied the mass of printed material that subsequently was published for analysis of this famous fire. The principal conclusion was that the fire indeed began in the vicinity of the barn owned by Mr. and Mrs. Patrick O'Leary at 137 DeKoven Street (now the site of the Chicago Fire Department Training Academy), but whether a cow was to blame is open to question. After the fire, investigators decided there were several possibilities of the cause: ignition by a lantern; ignition by smoking materials; deliberate incendiarism; and, possibly, spontaneous ignition of hay in the barn. Whatever the reason, the fire began Sunday evening about 8:45 and spread quickly.

At the time, wind gusts did not exceed thirty miles per hour, but they were sufficient to move sparks and embers that were raised by thermal updraft. The wind, from the southwest, spread the blaze four and a half miles northeast and north of the O'Leary barn. Notification of the fire department was delayed, although a druggist on 12th street later stated he had turned in two separate alarms from a city fire alarm box attached to his store. Subsequently the fire department denied receiving any alarm from that box during the first fifteen-minute period. In those days keys to alarm boxes were given only to reliable citizens, because false alarms were extremely prevalent.

Scene on Dearborn Street in Chicago as fire fighters and civilians react to the dire emergency. Note ground ladders at right, steamers back center.

Actually, the fire department responded to a still alarm given to Engine Company 6 by its own man on the watch desk, and the company, guided by the intense glare of the flames, arrived at a position behind the fire about seventeen minutes after nine o'clock. If they had taken a different position it is possible that the fire might have been checked, but these men were exhausted from their duty the day and night before. The best they could do on arrival was to protect the houses of the O'Leary family and neighbors, but they could not stop the spread of the fire.

The next alarm was given by a lookout on top of the court house tower who noticed the light of the fire at 9:28. At first he misjudged its location, but Engine Company No. 5, already responding to the alarm, saw the flames and went to a position in front of the fire on Taylor Street. Unfortunately, Steamer No. 5 became disabled and the fire fighters had to remove coal from the box and make repairs. Then they discovered that there was no coal, and someone had to race back to the engine house for more. Meantime, members of the crew used broken boards from sidewalks and fences to try to develop sufficient steam power for the pumper. This delay allowed the fire to cross to the north side of Taylor Street, and from there on it raced unchecked.

Next, the fire department tried to make a stand near the south bridge of the Chicago River, but by the time the fire reached this point all apparatus had been committed to the West Division. Very quickly, flying brands ignited structures in the South Division, including the local gas works, and soon the gas receivers ruptured, adding fuel to the flames. Then the fire passed through the business district, crossed the river into the North Division and, at 3:20 a.m., October 9, involved the wood shingle roof of the city water works. Here steam engines were pumping water from a tunnel running two miles out into the lake, and this water was supplying hydrants used by the fire department. At 4:00 a.m. those engines had to be shut down, and the only water available to the fire department was what could be drafted from the river and lake. Fortunately, on the evening of October 9 sufficient rain arrived to help fire fighters finally control the spread of flames.

About 2,000 of the 23,000 acres in Chicago were burned and 17,500 buildings out of 59,500 were destroyed with loss estimated at $200 million. Some 250 to 300 people were killed although exact counts of the fatalities were never accomplished. 98,000 people lost their homes in the conflagration. Martial law in the city was established under command of Lieutenant General Philip Sheridan

Thousands of people passed the Chicago Chamber of Commerce building and other involved structures. About 300 people lost their lives and more than 1,700 buildings were destroyed in this huge conflagration.

(late of the Union Army) commanding the Northwest Military Department.

Despite the severity of this disaster, within two weeks merchants were building scores of one-story sheds which quickly increased to several thousand; then new permanent structures of granite block or brick were started. Even so, many people lost all their possessions and sustained acute hardships as result of the fire. Of the $200 million loss, only $88 million was insured; and because of multiple bankruptcies of insurance companies only $45 million of that sum was eventually paid to claimants.

A year later, the famous periodical of that era, *Harpers Weekly*, made this editorial comment about the disaster:

The total losses by the great fire in Chicago, October, 1871, amounted to $200,000,000 to which another million must be added on account of the depreciation of property and the interruption of trade. The year which has passed since this event has seen at least one-third of the value of the destroyed property restored. The hotels, the places of amusement, the warehouses, the churches, and the schools which have taken the place of those which were destroyed are grander and more substantial edifices, and architecturally more beautiful. The prices of real estate are higher than at the time of the fire, and the industrial interests of Chicago have been more than re-established. In fact, the great disaster of last year is beginning to be regarded as a blessing in disguise, and the great Western metropolis — already connected with the interior by a score of railways, and having a lake marine rivaling the tonnage of the great sea-ports of the world — dreams with unabated enthusiasm of ship-canals westward to the Mississippi and eastward to the sea-board.[8]

Sketch of a family in Chicago about to be victims of the tremendous heat and flames.

Sketch of typical dwelling near Peshtigo, Wisconsin indicates heavy involvement of timber and brush by the racing tornado of fire.

Peshtigo — Engulfed by a Forest Fire[9]

On the same day that Chicago, Illinois, was devastated by conflagration, the small, lumber community of Peshtigo, Wisconsin, suffered one of the most severe loss-of-life fires in United States history. On October 8, 1871, a violent fire storm developed from a tremendous fire in surrounding forests, swept into the tiny community. Within the next twenty-four hours nearly 800 persons died and every building in Peshtigo (except one home under construction) was destroyed. This was a "classic" example of how quickly unchecked forest fire can destroy a community of combustible construction and meager fire protection.

It was several days before the tragedy became known to the nation because of the great publicity given to the Chicago fire. Peshtigo was isolated in the woods of northern Wisconsin, and it was not until the day following the fire that neighboring villages were notified and help began to arrive.

Like the people in Chicago, the 1,750 residents of Peshtigo had experienced the dry summer and fall but they had no sense of danger immediately before the fire. The village was surrounded by woods and fires were a common experience. During the summer, two fires had started outside the village but inhabitants had prevented them from reaching the buildings. There was a great deal of thick-barked, mature timber in the woods, and this usually was not harmed by ordinary fires, which became more of a nuisance than a threat.

On the evening of October 8, about 9:30, most of the people in Peshtigo were preparing for bed. At that time, through the smoke south of the village, a dull red glare could be seen near the horizon. A bit later, a breeze developed from the southwest and the air became very hot. Then a low rumbling noise was heard, like a train approaching from a distance. The sound increased to a roar and the threat of the fire heading toward the village caused mothers to take children from beds and dress them while the men prepared for fire fighting. At about ten o'clock flames could be seen at treetops near the village.

When a fire "crowns" in treetops it can spread very rapidly, then drop to the ground to start more intensive fire. This is a bad situation because of the additional involvement of fuel and the strong thermal updraft which carries sparks and brands along to ignite other fires. This is what happened at Peshtigo.

As a lumbering town, Peshtigo not only had wooden sidewalks but also applied a great deal of wood sawdust on the roads to keep down the dust. Shortly after ten o'clock, burning brands from the trees began to ignite the sawdust, the plank sidewalks and the combustible roofs, and so many fires started that local fire fighting forces could not control them.

The people could not flee from the fire — it was moving too fast. Some rushed into large buildings and soon were burned to death. Others ran to the river where many drowned, but a few survived. Three persons sought protection in a large water tank at a sawmill, but the water became so hot that all three were found dead the next morning. Others reached a low marshy piece of ground on the east side of the river, and survived because the standing hay in the marsh had been burned in one of the previous fires.

The following day survivors returned to the village and found one building standing amid ashes, pieces of iron and crockery, and the brick dry kilns of a woodenware factory. The disastrous blaze covered a stretch of some sixty miles north and south, and twenty miles east and west. In only a small part of that area did the fire travel for extensive distance through treetops. Subsequently, survivors described the thermal updrafts of the Peshtigo fire as "tornado-like." Others reported seeing balls of fire appear out of nowhere and suddenly disappear, apparently as hot gases struck a supply of oxygen.

Today, the City of Peshtigo, on the border of Wisconsin approximately 250 miles north of Chicago, is a modern community of 3,500 people. It is still ringed by farms and forest lands and has a number of small industries. It has changed considerably in the past 100 years but a special cemetery marks the unforgettable fire tragedy. (See plaque.)[10]

Many Peshtigo citizens tried to escape the fire storm by leaping into the nearby river but some did not survive. Nearly 800 persons were fatalities in that blaze and every dwelling in the town was burned to the ground.

Currier and Ives sketch of the great fire in Boston, November 9 and 10, 1872. Fire began on Saturday evening, lasted for fifteen hours, destroyed over sixty acres of buildings and caused property damage estimated at nearly one hundred million dollars. (Library of Congress Photo.)

The Great Boston Fire of 1872[11]

In November, 1872, approximately a square mile of the business district of Boston was wiped out by a conflagration.

A study of reports and records of the Boston Fire Department of 1872 shows a frightening list of deficiencies in the municipal fire protection of that time. Fortunately, the city learned its lesson and the situation bears little resemblance to Boston of today.

The fire department had 472 members of which only eighty-nine were permanent men; the others were paid only for duty at fires. Chief Engineer John S. Damrell was paid $3,000 a year and was elected annually by the City Council along with eleven district engineers who were call men. The engineers, stokers, and drivers of steamers were permanent, as were the drivers of ladder trucks. All firemen were call men, most of whom were paid $300 a year.

There were twenty-one engines, ten hose and seven ladder companies, but only six engines, five hose and two ladder companies were located in the city proper. Fire companies elected their own officers annually. Most of the downtown steamers were rated at 500 gpm "when working at a fair speed," but several steamers in outlying areas were as small as 250 or 300 gpm. Steamers were normally pulled by two horses, but at the time of the fire, most of the horses were sick with epizootic disease (equine influenza) and apparatus had to be drawn by hand. Very few engine companies had separate hose carriages. A two-wheel reel with 300 to 500 feet of 2½-inch leather hose was towed by the steamer. Of course, this added weight slowed the response. Steam was not maintained in the engine boilers, as was a later practice, and quick-hitch harnesses had not been devised. This probably did not matter, as fire fighting depended on call men who responded from other jobs when the tower bells rang.

Each hose company carried from 500 to 1,000 feet of "leading hose" on a one-horse carriage.

Chief Damrell had complained to the Water Board for several years, without success, of the inadequacy of the water supplies. These were furnished chiefly by 4-inch mains. Hydrants were supplied by 3-inch branch pipes. The Water Board did not share the Chief's concern. Fortunately, a number of cisterns (also fed by 4-inch mains) were maintained from which engines could take suction. Chief Damrell testified that it was impossible to concentrate streams at any one location, because one engine would take all the water from any main.

Numerous witnesses testified that during the conflagration streams of water did not go above the third floor level, although 5- and 6-story buildings were common in the fire area. There were few building regulations, and fire-resistive construction was unknown. Instead, wooden trim and dormer windows were common on the upper floors. Exposure protection was almost unknown, and fire crossed the narrow streets above the heads of the firemen. Automatic sprinklers had not been invented.

There were no water towers or ladder pipes or other means of getting streams to upper floors. Aerial ladders and water towers had not been invented. Buildings did not have fire standpipes. The longest fire ladder was forty feet. Modern extension ladders had not been invented. Short ladders could be spliced to reach 62 feet, but it took a dozen men using poles to raise the spliced ladder. There were no heavy stream devices.

Help came from a total of thirty outside fire departments from as far away as New Haven, Connecticut; Portsmouth, New Hampshire; Biddeford, Maine; and Providence, Rhode Island. Help included 1,689 fire fighters with forty-five engines, fifty-two hose carriages with 41,050 feet of hose, and three ladder trucks. Hose threads would not connect to Boston hydrants, so outside engines drafted from fire cisterns.

Subsequently it was determined that the conflagration destroyed 776 buildings, took thirteen lives and caused property losses of about 75 million dollars. Two fire fighters, a foreman and an assistant foreman were among the fatalities.

In those days communications left much to be desired. The telephone had not been invented.

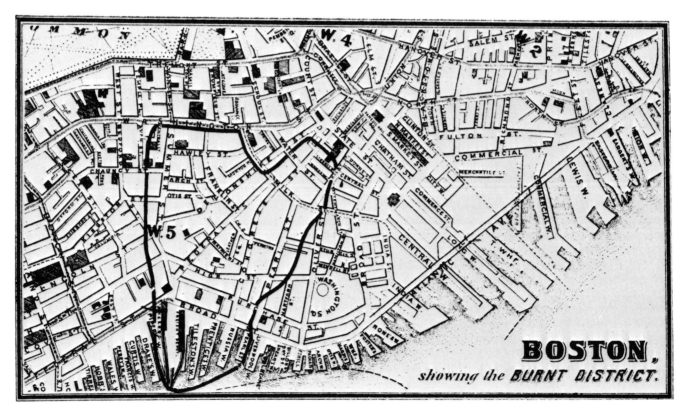

BOSTON, showing the BURNT DISTRICT.

They did have telegraph which was used to call outside help. Within the department, alarms were given over the fire alarm telegraph. There were street boxes, but these were kept locked with keys in the custody of reliable citizens. The locks had to be changed from time to time to prevent false alarms. The fire alarm system was not under the control of the fire department. There was no automatic fire detection equipment in buildings. The fire was visible several miles away before the first alarm was sounded. Not only was the building of origin fully involved but three major exposures were well alight when the chief arrived within 10 minutes of the first alarm.

There was no organized fire prevention program within the fire department and no personnel assigned to enforce the meager fire prevention regulations.

As a result of hearings held after the fire, the department was reorganized and placed under a Board of Fire Commissioners. Within a relatively short time a great many improvements were made. All companies in high value areas were manned with full-time permanent fire fighters. A number of new companies including a fireboat company were placed in service. The fire alarm system was transferred to the fire department. A training program was started. A fire department maintenance shop was provided. District chiefs were made permanent and given responsibility to familiarize themselves with buildings in their districts for pre-fire planning. Fire prevention and building regulations were improved. Improvements were also made in the water distribution system and additional hydrants were installed on larger pipes.

The city of Boston had received an unforgettable lesson of the potential of fire.

The Illustrated Police News also showed this action. Since the fire department's horses were sick the steamers had to be pulled by hand. Hose cart at right otherwise would have been attached to rear of steamer.

Here is how the U. S. Patent Office looked before the 1877 fire. Today the Patent Office is in a handsome new building in Crystal City, Virginia.

The Great Patent Office Fire of 1877[12]

When the average citizen views the imposing landmarks of American cities he seldom realizes that many of these great structures would not stand today but for the valor of America's fire fighters down through the years. A typical incident was the great old Patent Office Building, covering two city blocks in downtown Washington, D. C., more recently housing the U. S. Civil Service Commission. Like many other great structures, this notable public building had its ordeal by fire many years ago.

Congress first authorized the erection of the Patent Office in 1836 and that part which faces on F Street, Washington's main shopping thoroughfare, was completed in 1840, more than one hundred and twenty-five years ago. Its well-known south portico was an exact copy of the famed Parthenon in Athens. The 7th Street side was completed in 1852; the 9th Street side, where the great fire started, was completed in 1856. The rear on G Street side was finished in 1867. The building cost $3,000,000, a very large sum in those early days. It was 403 feet long and 274 feet wide, north to south.

The fire of 1877 destroyed thousands of models of early inventions by American inventors. At that time a model had to be submitted for each invention, in addition to the usual plans and specifications. The fire also destroyed countless valuable records, as it occurred long before the establishment of the National Archives became a depository for official documents and records.

When the fire occurred, the District of Columbia Fire Department consisted of only six steam fire engines and one ladder truck. However, a considerable force was assembled to fight this fire. Through the workings of mutual aid, help was obtained from Baltimore, Maryland, and from the U. S. Navy Yard in Washington. One account says that horse drawn apparatus also responded over the road from the Alexandria, Virginia, Volunteer Fire Department.

The *Firemen's Record*, published almost a hundred years ago in Washington, D.C., gives a graphic account of how this big fire was handled:

At 11:30 a.m., September 24, 1877, an alarm was received from box 131 (then located at the corner of 12th and G Streets, N.W.). The first alarm engines responded with their usual promptness; what was their surprise when they saw that the Patent Office was on fire. The Chief Engineer arrived on the ground in about five minutes after sounding of the first alarm. Upon his arrival he at once surmised that the fire had originated under the roof, about the middle of the West or Ninth Street wing. He immediately summoned the remaining companies in the city as the fire had gained great headway.

At this time the fire was confined to the 9th Street wing in what may be termed the loft or attic of the building, a space between the ceiling of the third or museum floor and the roof. The space was about 12 feet high and sloped to about 4 feet to the sides. It was stored with a very large quantity of inflammable matter, to which was added the dry pine rafters and sheathing which supported the copper roof, making it an easy prey to the flames.

This attic, like the large halls beneath it, traversed the entire building; in other words, at no point were there partition walls or other impediment to stay the progress of the fire. This immense area, with its low ceiling, of necessity created a terrible draft and notwithstanding the wind was from the south, the fire burned in every direction, but with greater rapidity towards the south, or directly against the wind, reaching the F Street wing some 20 minutes before it did the G Street wing.

A company was at once ordered to the roof, when just as they gained it the discovery was made that there was but one outlet from it. This was on the F Street front over the main portico reached immediately from a room used for storing specifications and drawings of which there were nearly 8,000,000, aggregating in weight over 200 tons. The stairway being a narrow wooden concern liable to ignite from the heat at any moment.

The room spoken of was reached by double flights of fancy iron stairs rising from the museum. A glance showed that this stairway must be saved, as it was the only means of retreat for the men then on the roof, and that in order to save the stairs the fire must not be permitted to get into the F Street wing.

To this end four of the companies — 1, 2, 3, 4, as fast as they arrived were placed at the point where the 9th and F Street wings unite. Their efforts were not in vain as the fire was checked at this point.

At this time all the possible apparatus, including that from the Washington Navy Yard, were at work and it was deemed advisable to ask for assistance from Baltimore. Shortly after 12 o'clock Mr. Thos. M. Byrne, one of the Washington Fire Commissioners, telegraphed to Baltimore for aid. Upon receiving the dispatch in Baltimore, Box 13, which was located at B. & O. R.R. depot, was sounded, and the firemen responded promptly. They were informed of the call from Washington and no time was lost in getting quickly the apparatus on board the cars.

Engines 1 and 2, with their hose carriages and an abundance of hose were detailed. At 1:05 the train moved amid the cheers of a multitude of citizens who were congregated around the depot. The firemen reached Washington shortly after two o'clock and were received by cheer after cheer of the citizens. The apparatus immediately proceeded to the scene of the fire, arriving in ample time to render valuable assistance.

A second dispatch was sent for more engines and two more companies were detailed to go; Engine Companies 3 and 4, left at 1:45. This train arrived at 2:41 p.m., having made the run of forty miles in fifty-four minutes. At every station on the way the firemen were greeted with cheers.

When the train arrived, cheer after cheer greeted the firemen as they disembarked. Companies 1 and 2 took up their positions at the corner of 6th and H Streets, and 5th and H respectively, and were in time to go to work. Previous to the arrival of engines 3 and 4, while the Patent Office was still burning, another threatening fire broke out on G Street. The Patent Office demanded the attention of our own department. It was a critical moment, for it seemed certain that a vast amount of property would be destroyed. Engines 1 and 2 from Baltimore came to the scene and prevented the fire from spreading. It was about this time that companies 3 and 4 arrived, and the fire, with their assistance, was soon under control.

As the Baltimore firemen were getting ready to go home, Secretary Carl Schultz sent for Chief Hennick and requested him to bring his men into his office, so that he could thank them personally. The firemen soon assembled: the Chief introduced the Secretary, who said: "Gentlemen of the Baltimore Fire Department, it is my duty and pleasure in the name of the Government to thank you for the valuable aid that you have rendered us today. You probably are not aware how great the service you have performed. Had it not been for your aid, many records of incalculable value to this department would have been lost. When we called you, we did not call in vain, you came with the speed of lightning and in the nick of time, and rendered inestimable service both to your country and the City of Washington, and for which you will ever have our thanks.

The firemen gave three cheers for Secretary Schultz, and the Washington department. In return, cheers were given with a will for the Baltimore firemen. The firemen then took their places on the apparatus and were driven through a number of streets, and were received with cheers from the citizens. After parading the principal streets the firemen arrived at the depot, where the last good-bye was said, and the embarkation for Baltimore took place.

Necessarily this fire in a great government building, created the most intense excitement, and at a very early period, after its discovery, not only the streets in the vicinity, but the halls and corridors were thronged with citizens and employees, each anxious to render such assistance as he could; but as is usual in such cases, none doing any real service for the reason that there was no unity of purpose or action.

The officers and members of the department, without exception, maintained their well earned reputation for courage, judgment and skill during the incessant and arduous labor of the day. They were ever at their posts, facing danger unflinchingly, and demonstrated by their actions that they could, at all times, be relied upon to do their whole duty.

Other Major Fires

In Colorado Springs, Colorado, October 1, 1898, a fire started in the freight station of the railroad at a time when gale winds were blowing at forty-seven miles per hour. Within five minutes the entire freight station was engulfed by flame and very quickly ten freight cars were also involved. One of them contained a carload of dynamite and another had black powder. A train crew was able to uncouple the car with dynamite and pull it away from the fire but they were unable to reach the car containing the explosive powder. Mutual aid had been called from three volunteer fire companies from Colorado City and Manitou, but despite aggressive fire fighting, it was realized that the explosion was inevitable and fire fighters were withdrawn to avoid casualties. Just one half hour after the first alarm the powder blew, throwing brands and embers high into gale winds, where they were carried northeast to land among a group of twenty frame dwellings and three lumber yards.

Very quickly the fire turned into a conflagration and fire fighters at some positions had to abandon hoselines and equipment. Subsequently additional mutual aid was received from Denver and Pueblo. One chief officer realized that the hose coupling threads of Denver and Colorado Springs did not match so he arranged to pick up 2,400 feet of extra hose from fire department shops and as many hose jackets as he could obtain. (A hose jacket is used to clamp around a small leak in a fire hose; also to join sections of hose together when coupling threads do not match. Six years later during the famous Baltimore conflagration of 1904 this same problem of nonmatching hose threads was to be a major factor.)

Eventually, the powerful wind gusts died down but not until the famous Antlers Hotel had been completely destroyed, along with other small properties. It was estimated that at least 200 roof fires occurred, most put out quickly by householders using garden hose or buckets of water.

The factors that contributed to this conflagration were identified as: high winds; many combustible buildings with no fire walls, hazardous occupancies, wood shingled roofs, easily ignited; and lack of fire pumps and equipment that might have provided effective protection of exposed buildings.[13]

In the last three decades of the 1800s, New York City suffered a great deal of political turmoil and a shocking series of major fires. Immediately after the draft riots the fire department became a paid organization named the Metropolitan Fire Department but the volunteer members, who had served the city in the colorful, brawling, violent years since the turn of the century, did not want to give up their duties very quickly. For a long time, the fire fighters operating in the professional fire department were victims of stoning, rough treatment and other handicaps as the volunteers and their political supporters showed their anger. The new paid fire department was to have 583 men including 100 to man the hand engines.[14]

A year later two other major fires occurred at a time when many fires of suspicious origin were happening all over the city. Within a twelve-month period there were forty-six outbreaks of fire, forty-five arrests for arson and fourteen convictions. Two of the more serious incidents involved the Academy of Music, burned to destruction May 21, 1866, and the Chittendon Building fire on Broadway the following year, which brought more organized opposition to the new professional fire department. In the Academy fire, two fire fighters were killed and a third seriously injured when gas, accumulated from the theater lighting system, suddenly exploded.

In February, 1876 another overwhelming fire occurred on Broadway, involving twenty-two buildings at a loss estimated at $1,750,000, and killing three members of Engine Company 30 who were caught beneath a falling wall.

The most shocking disaster occurred across the East River in Brooklyn on December 5, 1876. In the Brooklyn Theater, one of the great showplaces in the city, a play was being presented with the popular Miss Kate Claxton in the leading role. The last act was nearly over when fire was discovered backstage. Nearly a thousand persons were packed into the auditorium and Miss Claxton continued to play her role until the stage actually became involved in flames, but about 500 spectators tried to dash in panic from the upper balcony. (Later it was determined that there was ample time for everyone to escape from the building, but the frightened people packed into the stairway, screaming and fighting in a struggling mass, became completely wedged and unable to move.) Very quickly, 296 men and women burned to death on the stairs.

After the fire more than 100 who could not be identified were buried in a common grave in Greenwood Cemetery.

This tragedy prompted the city to inspect all theaters and showplaces more rigidly to see that adequate exits were provided.

Two major building fires occurred in 1875, within three days. On January 4, a fire got out of control on Broadway and destroyed buildings on that famous street intersecting Grand Street. Loss was over a million dollars and one fire fighter was killed. Three days later another fire destroyed seventeen buildings and forty-eight business firms.[15]

Two other loss-of-life fires occurred in the 1890s in Manhattan. On August 22, 1891, the Taylor Building on Park Row suddenly crashed to the ground burying dozens of people, followed by ignition of the debris. Fire fighters worked for hours pulling victims out of the ruins and the death toll was sixty-one. Six months later the Hotel Royal on 8th Avenue was destroyed by fire and twenty-eight people died in the building.

The Trend to Paid Fire Departments

At this time an important change was taking place in many of the larger communities, namely, the recognition that fire was a major threat and that a well-trained, well-equipped group of fire fighters had to be available for prompt response to every fire emergency. For nearly one hundred years volunteer fire companies of varying size and capabilities had rendered great service for the conflagrations, the explosions, and the other major disasters, but the high frequency of arson fires and the many minor incidents of injuries from burns and other fire damage, indicated the need for fire department manpower that could be available at any time, day and night. This need, coupled with the trend in fire apparatus that required better trained engineers and drivers, persuaded the people in many communities to request full-time paid fire departments.

As this trend developed it was not by complete acclamation. In many states and communities, volunteer fire departments had become highly important to the social and political fabric of the democratic process, and change to a tax-supported fire department simply would not be accepted. Even today, the assessment of the role of the fire department in a given community is argued on the same premises that caused violent controversy a hundred years ago.

In 1975 the number of volunteer fire departments or independent fire companies amounts to about ten times the number of paid fire departments and the amount of personnel in each group seems to fall within the same ten-to-one ratio. In the industrialized states, there is a larger quantity of the paid group; but in other states, the volunteer Fire Service has functioned as a cohesive force since the eighteenth century. In Maryland, Connecticut, New Jersey, Pennsylvania, New York, Virginia, the Carolinas and other states the volunteer fire companies can trace their heritage back to Colonial times, to the Revolutionary War, or to the early part of the nineteenth century, and their pride in past traditions and accomplishments is an important factor in how they are organized and how they fulfill their operational standards today.

But in the larger communities, the changeover to a professional fire fighting force was a necessity and the following summary indicates how this developed in some of the major cities:

Like many others, the fire department in Newark, New Jersey, made gradual change from volunteer to paid basis. In 1860, there were thirteen fire companies in service; two had steam engines, horse drawn; the rest had manual pumps. The Exempt Association, with a membership of 127 arranged to purchase a first class Amoskeag steamer named the "Minnehaha." This made a large difference in operations, especially when it was supplemented by a second class steamer whose delivery had been delayed by exhibition at the U. S. Fair of Agriculture in Cincinnati, Ohio. With these two powerful engines the department had to consider the possibility of permanent manpower, but it was not until June 1, 1889, that Newark became established as a fully paid fire department.[16]

In California, the city council of Los Angeles had the power to create a fire department after incorporation in 1850, but this power was not exercised; for the next twenty years the growing city was protected by a volunteer brigade that used three-gallon leather buckets for fire fighting. In 1870, a meeting was called for ex-members of the San Francisco Fire Department and members of other fire departments who lived in or near Los Angeles. After formal discussion and procedures, the first Los Angeles volunteer fire department was formed in September 1871, designated as Engine Company 1. Apparatus consisted of a hand-drawn Amoskeag pumper and a hose cart. The Amoskeag had been manufactured in Massachusetts and shipped to San Francisco by rail; from there it was brought overland to Los Angeles, drawn by mules.

In 1874 members of Engine Company No. 1 asked the city council to purchase horses for pulling the heavy apparatus, but when this request was denied the company disbanded. A few months later the company was reorganized and a second engine company was formed, known as "Confidence Engine Company."

In 1875 a local manufacturing company built the city's first hook and ladder apparatus. This was used for a short time but proved to be too heavy and awkward, so it was sold to the town of Wilmington. A year later the formally organized volunteer fire department of Los Angeles was established with Charles E. Miles elected as chief engineer. In the same year the village bought a second hook and ladder truck and a 65-foot extension ladder.

Representation of fireground action during the great conflagration of 1866 which swept through the City of Portland, Maine. It burned half the city, left 2,000 persons homeless, destroyed every bank, a newspaper office, and city and county buildings on Congress Street. It was started by a firecracker tossed into wood shavings in a cooper's shop.

Three or four hose companies with volunteer membership were organized between 1878 and 1883 and rather quickly a very fierce rivalry developed between all these companies. "Taking a hydrant" became a maximum objective in each fire alarm response and violent fights broke out as each fire company tried to be "first in." It was not long before certain individuals were designated to be "plug guards" to save hydrants for first in companies. The violence of fireground encounters was not tolerated long by the City Council. On January 12, 1886, the Council created the paid fire department, at a time when the volunteer companies included 380 members. By 1889 the fire department had thirteen steam fire engines, eleven in service, two in reserve. There were also twelve combination chemical and hose wagons, ten hose wagons and hose carts, four hook and ladder trucks, four chiefs' buggies, two supply wagons and eighty horses. Three more steam engines with accompanying hose carts and two Champion chemical engines were purchased in 1890. Five years later the department added three more Champions and three hose wagons. In 1897, the city purchased its first 85-foot aerial ladder, made by the Babcock Company.[17]

In Savannah, Georgia, the first volunteer fire company of twenty-one members was appointed by an act of the general assembly but it was not until February 1, 1890, that the call force of the department was abolished and the department went on full-time, paid basis.

In Charleston, South Carolina, volunteer fire companies went out of service January 1, 1882, when the paid fire department was formed.

The Omaha, Nebraska Fire Department, formed as a volunteer group in 1860, had its first fire apparatus built locally — a handpulled cart with hooks on the sides to hold ladders and buckets. It was about twelve feet long, painted bright red and had small wheels for turning corners easily. The ladders were made from tall trees that grew on bluffs north of the city.

In 1867 the city bought a third-class Silsby rotary steam engine, followed by a second steamer, purchased from the Ahrens Company in Cincinnati, Ohio in 1869. A year later another Silsby, named the "Nebraska" was purchased, just like the first one. Omaha began its paid fire department in May, 1885.[18]

The San Antonio, Texas Fire Department, by now fourteen years old, bought a hand-drawn steam pumper in 1868 but did not have the use of substantial municipal water supply until ten years later. By March 1, 1891, the department was fully paid with a roster of thirty-five men — seven per engine company and seven per ladder company. It had a central alarm system and a budget of $25,000 annually.[19]

The Portland, Oregon Fire Department had been organized in 1851 as the Pioneer Fire Company Number 1. Its first hand-drawn pumper was built by a local blacksmith. Portland got its first steamer in 1856 and added a Silsby and an Amoskeag in 1883 when the department became fully paid. Its first big fire was in December 1872 when three business blocks were destroyed, but just about a year later twenty-two blocks were demolished in another conflagration.[20]

The Dallas, Texas Fire Department was formed as Fire Company Number 1 on July 4, 1872, reorganized as a paid department in 1885 with horse drawn apparatus, and accepted its first motorized apparatus in 1899.[21]

The San Francisco, California Fire Department went from volunteer to part-paid status December 3, 1866, but early in 1900 the city adopted a new charter, replacing 344 call men with paid personnel. The fire department had its first fire boat in 1878, a converted tugboat eighty-six feet long, carrying 1200 feet of hose and capable of pumping 65,000 gallons per hour. The first chemical company was formed May 18, 1890 and the first water tower went in service in 1891.[22]

The Honolulu, Hawaii Fire Department was using horses in 1886 but did not receive its first steamer until 1891.

In that city, on April 19, 1886, a major fire started in the Chinatown district when two men argued over lottery tickets. When one man tried to burn the tickets, nearby combustibles ignited and the fire subsequently involved sixty acres of buildings. King Kalakaua worked at this blaze and a month later he established and signed Honolulu's first building laws.[23]

Uniform of typical fire chief in late 19th century. This is Chief Richard Ardagh of Toronto, Ontario who died of injuries received at a fire in 1895. Note ornate speaking trumpet, white helmet and formal dress uniform.

Trends in Fire Apparatus

From shortly before the Civil War until the end of the 19th Century competition among builders of fire apparatus was intense, and many firms started their manufacturing, delivered a few steamers, then went out of business. Amoskeag, Babcock and other famous names had already made their mark in history. Smaller, less successful builders began and failed.

In South Carolina, the Old Charleston Fire Company of Axemen purchased its first steam fire engine in 1861 and two new fire companies were formed in 1865 and 1866 to operate other newly-purchased steamers.

Boston tried out its first self-propelled Amoskeag steam fire engine in 1867 and the fire department report stated "it was observed that it did not frighten passing horses any more than did the horse drawn steamers."

New York had its first water tower in service in 1880.

Out in the kingdom of Hawaii, the Honolulu Fire Department during the reign of King Kalakaua received its first horse drawn steamer in August 1891. Merchants of the town purchased a pair of horses trained especially for this apparatus, then equipped a station house with electrical appliances for turning lights on automatically and releasing the horses from their stalls. A month later the department received a Champion chemical engine with two eighty-gallon tanks.[24]

It is worth noting that the Honolulu Fire Department is the only one in the U.S. and Canada that had authentic kings serving as active members. In the 1850s King Kamehameha took an active interest in the department, and in the 1880s King Kalakua was the first secretary of Honolulu's No. 4 Engine Company.

Some of today's apparatus builders started their manufacturing firms in those years. In Detroit, Michigan in 1881 F. S. Seagrave founded his "Seagrave and Company" firm to build ladders for fruit orchards; but a short time later local volunteer fire fighters asked him to build a vehicle to transport ladders to fires. The first Seagrave ladder was mounted on two wheels, later improved to a four-wheel appliance, designed to be pulled by the volunteer. Next to be developed was one of the first chemical fire trucks with two reels of hose, oil lamps, warning bells and an extension ladder. The driver and one man rode in the front seat back of the team of horses and two other fire fighters hung onto the rail at the rear.

In 1891 Seagrave moved his business to Columbus, Ohio and changed the name of the firm to "The Seagrave Company."[25]

One of the horse-operated piston pumpers built and distributed by the Howe Company in 1875. (Howe Fire Apparatus Photo.)

The Howe Fire Apparatus Company was started by B. J. C. Howe in Indianapolis, Indiana, later moved to Anderson, Indiana. Howe introduced the first fire department piston pump that could be operated by a team of twenty men, or a team of horses that pulled it to the fire scene. The company, which has remained a family concern up to the present, later made a variety of horse-drawn and self-propelled apparatus, but in those early days pumps for town water supply received a lot of public attention. L.M. Howe, a son of the company founder, on holidays had the task of inspecting all horse watering troughs in Indianapolis to make sure that the well pumps were in shape to handle the heavy demands of thirsty horses. At that time the Howe Company was supplying piston pumps for the troughs, and also was making horse-drawn pumpers for the Indianapolis fire stations. L.M. Howe also had to check these to make sure they were up to performance for the inevitable fire fighters' "water battles" of the holidays.

In 1904 when Henry Ford, Elwood Haynes and others were starting the first automobiles, the Howe Company began experimenting with pumps on automobiles and had its first practical automotive pumper ready in 1908. But in those early years this type of pumper was more of a novelty than an efficient machine and fire departments were slow to accept the change. The company's "Baroda" pumper, horse-drawn, with a motor-powered pump, was popular from 1901 to 1914.[26]

The Waterous Company of Saint Paul, Minnesota, also began as a family organization. In 1844 the Waterous Engine Works Company, Limited, was started by C.H. Waterous to build different types of machinery, including fire apparatus. Because the company grew rapidly, he sent his twin sons, Frank J. and Fred L. Waterous, to Winnipeg to open a branch office. Five years later a new plant was opened in St. Paul, Minnesota. In that same year the company introduced its first steam-powered pumper, to be horse-drawn or hand-drawn, followed by its first gasoline-powered pumper in 1898. By 1906 Waterous developed a self-propelled apparatus with a dual engine, one for pumping, one for propulsion, of a type purchased by several large fire departments. The company also produced aerial ladder trucks, chemical engines, and later, motorized apparatus.[27]

A Silsby horse drawn engine in Washington, D.C., 1885. (Smithsonian Institution.)

First and only self-propelled Amoskeag steam fire engine in Detroit, Michigan. Called "Hercules", it was placed in service in January, 1874. (Clarence C. Woodard Photo.)

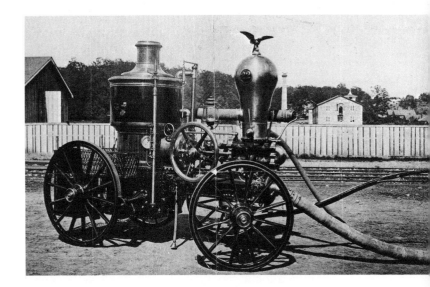

This was the "Lily of the Swamp" steamer in Frederick, Maryland in 1878. It was built by the Clapp and Jones Company. (Smithsonian Institution.)

A Poole-Hunt engine in Baltimore, 1870. (Smithsonian Institution.)

For more than fifty years American fire departments relied on horses to pull apparatus. A three-horse hitch like this always attracted attention from spectators like those at right.

That Wonderful Era of Horses

It is difficult to identify the exact date when horses were first used to pull fire apparatus, but apparently the credit for this innovation belongs to the Good Intent Fire Company of Philadelphia, Pennsylvania, which used horses in 1803. This was resisted strenuously by other organized volunteer fire companies in the city; such opposition was obvious in most cities and towns for almost the next fifty years. Actually, the Philadelphia Fire Department did not make full use of the capabilities of trained horses until 1858 when the heavy steam fire engines arrived. The fire department in New York City started using these animals in 1835 but opposition from the volunteers was as violent as in Philadelphia.[1]

Full acceptance of horses in the Fire Service did not occur until fire department pumpers powered by steam pressure became a practical reality; even so, the volunteers were more willing to test their strength against the steam fire engines than against the obedient horses. Once the steam pumper proved capable of sustained performance on the fireground, it was only a matter of a short time before the volunteers realized they could not match such performance by manual effort. The other reality was that such engines were of tremendous weight, some were five to eight tons, and hauling them to a fire called for an exertion of manpower that was not always present at the time of an alarm. A third reality was that horses were intelligent, affectionate, easily trained animals who quickly adjusted to the excitement and demands of fire department action, and, as each fire department or fire company began using horses, the volunteers as well as the paid fire fighters soon adopted them as personal pets. For a period of seventy to eighty years, in most of the larger fire departments throughout the country, the obedience, de-

100

votion to duty, and strenuous sacrifice of these animals became one of the great legends in the Fire Service.

Not all fire departments or volunteer fire companies used horses. Many smaller towns continued to rely upon light hose wagons and manual pumps pulled to the fireground by members of the local fire company. Some fire departments went directly from handdrawn apparatus into the motorized era, without ever utilizing the strength of horses. But in the larger cities, the tremendous task of hauling heavy apparatus over cobblestoned streets, or dirt roads, simply could not be performed efficiently by manpower. Horses did this work, and performed well, although during rain and snowstorms they often had to be helped by human effort.

At first, dray horses, or any other kind of strong horses, were used by the fire departments. It took some years before calculations were made of the best methods of fastening a harness, positioning the harness near the apparatus, placing animals in their stalls, attaching the harness and apparatus, and otherwise preparing for quick departure from a fire station. Decades passed before such minor operations were accomplished efficiently; meantime, the horses patiently adjusted to alarm signals and the actions of a great number of men in the respective fire stations.

Legends concerning the splendid animals passed quickly from one fire department to another, and news stories and historical data seem to verify the splendid behavior of horses in that very strenuous era of fire fighting. They hauled every type of wheeled vehicle in emergency response: chiefs' wagons, pumpers, hose wagons, aerial ladders, tanks, rescue vehicles, and anything else that was needed for an emergency. The length of time of their runs to alarm sites within their districts was remarkably consistent if their drivers were familiar with the route of response. The horses quickly learned what would happen when they arrived at the fire, how they would be disconnected from the apparatus and led to a safe spot until needed for return to quarters. Their greatest excitement was during the immediate action of responding to an alarm, and then the race to the fireground; the final hitching and trot back to the station was accepted in a quieter mood.

It is interesting to review comments in the histories of fire departments of nearly a century ago. For example, in 1887 the Los Angeles Fire Department had thirteen steam fire pumpers, twelve combination chemical and hose wagons, ten hose wagons and hose carts, four hook-and-ladder trucks, four chiefs' buggies, two supply wagons and eighty horses to haul this apparatus back and forth.

Fireman W. E. Hurst checks harness of two-horse team in Devil's Lake, North Dakota fire station in 1909. (Donald E. Gilman, NDFA.)

A 900 gpm American LaFrance steamer used by the Boise, Idaho Fire Department in 1910 had three-horse hitch.

Perhaps not speedy, but dependable, was this pair of oxen shown hitched to a Waterous single cylinder gas engine pump in Saskatoon, Saskatchewan. Photo taken in 1910. (Saskatoon Fire Department.)

In Montreal, Quebec, in 1919 this horse-drawn hydrant thawer provided steam to melt the ice and free the hydrant for operation.

Some of these vehicles, particularly the steamers, weighed well over four tons. The fire department report mentioned that horses were put through a three- to four-week training program during which they were taught how to back out of their stalls when an alarm struck. They had to wait in position until a harness was placed on them, then they moved to the front of their assigned piece of apparatus. These animals were so well trained for punctuality and were so eager that sometimes when an alarm came in they would leave the engine house before they were hooked up and before the fire fighters had time to get to the apparatus. This happened in many fire departments and it is easy to imagine the shouting and general confusion that occurred when the animals raced out of the station without their drivers![2]

Eventually, Los Angeles horses were purchased in teams of two, three, four, or more, usually of specific breed, strength, speed, color and other characteristics. Once they became part of a fire department, they were given individual names and their good and bad characteristics were quickly identified by the fire fighters.

The Omaha, Nebraska Fire Department had three named "Admiral Dewey," "Bill McKinley," and "Pete Smith," who lived in luxury on the apples and sugar lumps fed to them by the officers, fire fighters, and civilian visitors. In those days of unheated quarters, these horses were quickly covered by blankets during the cold-weather months.

Indication of such duty is contained in the "Prelude to Glory" part of the history of the Washington, DC Fire Department, which had this to say about hostlers, the men who cared for the horses in fire stations:

The hostlers shall have charge and due care of the horses of their respective companies while at the engine or truck house, going to a fire, whilst there, or returning to the house, and shall have them at all times ready for immediate service. They shall report to, and otherwise execute directions of the foreman, and shall be on duty at the engine, or truck house, at all times, except during fires; and whilst practicing horses shall not go farther from the house than directed by the foreman.

It was important for fire departments to keep these animals in excellent physical condition and they received the best of care and medical attention to assure this health. Nevertheless, at times, for the sake of economy and performance it was necessary for fire departments to get new horses. In 1872, in the District of Columbia, a campaign for "survival of the fittest" developed when the district fire commissioners ordered that additional

This is a famous picture of a New York steamer in April, 1915. Careful examination of the original photograph showed that all twelve hooves of the animals were off the ground when the picture was snapped — a bit of action many photographers tried to capture.

horses be purchased and only those that were "first class in every respect" were to be retained. In those days, a fire horse usually cost about $245, was over five years of age, and weighed something less than 1,500 pounds. Unfit horses might be retired to a city farm or otherwise would be destroyed and the fire fighters would try many ruses and delays to keep their pets from such a fate.[3]

In 1893 the Honolulu Fire Department purchased a pair of horses on May 1, and a second pair on July 18. From then on, horses were purchased regularly until all apparatus was hand-drawn. Eventually, Honolulu had a problem that was familiar to other fire departments: it was necessary to build stalls in fire stations for resting and bedding these animals. The stalls took up floor space and the hay and feed created something of a fire problem, but these minor inconveniences were accepted as necessities. Honolulu kept its horses until May 1920 when the last two were retired and replaced by a motorized pumper. These two pets, Jack and Jill (gray and black), were purchased by a local citizen for $660. Another

team, Bill and Jerry, which hauled the reserve chemical engine, were sold for $640.[4]

The San Francisco, California Fire Department made its first response with horse-drawn apparatus on August 19, 1863, when the Pennsylvania Engine Company No. 12 responded to its first alarm. Fifty-five years later, on August 21, 1921, the same company, now designated as Engine Company No. 6, responded to the last fire in which horses were used by the department.

The Portland, Oregon Fire Department converted two hand-drawn steamers (a Silsby and an Amoskeag) to horse-drawn apparatus in 1883, and the last horse of that department was retired in April 1920.

The Boston, Massachusetts Fire Department replaced its hand-drawn equipment in 1860, resorting then to horse-drawn steam pumpers and hose wagons. In 1867 Boston tried a self-propelled Amoskeag steamer but, since this did not move too well through the city streets, the department converted it to a horse-drawn wagon. Boston retired its last horse-drawn equipment in 1925.

A wonderful sight in New Haven, Connecticut about the turn of the century as Hook and Ladder No. 1 races down the street, accompanied by gleeful youngsters on the sidewalk. (From New Haven Fire Department.)

In 1909, the Baltimore, Maryland Fire Department had this handsome, white three-horse hitch named "Snag," "Germany," and "Kahl" after Chief Kahl. (Baltimore Fire Department.)

In the Newark Fire Department a veterinarian administers treatment to one of the horses fastened in a special restraining frame. (Newark Fire Department.)

When a box is pulled the first stroke opens the doors of the engine and hose houses and releases the horses from their stalls. Then the fire-force get the signal from the gongs at each house. By the time the box number is in the horses are hitched and ready to go. The big bells on the houses are nearly all electrically connected. The circuits are connected with an automatic repeater and thus the box numbers are repeated without the assistance of an operator. *From the history of the Indianapolis, Indiana Fire Department.*

"Dewey," "Schley," "Fitz," "Joe," "Nancy," and "Flora" were horses of the Phoenix, Arizona Fire Department starting service in 1899. The first two hauled the largest steam pumper in the department; Fitz and Joe hauled the hook and ladder; and Nancy and Flora pulled the hose wagon. The six of them rendered faithful service until the department adopted motorized apparatus in 1914.[5]

Sioux Falls, North Dakota purchased a horse-drawn Silsby steam pumper in 1903 and retained its horses until 1917, when the department became completely motorized.

The last run of fire horses in a downtown Detroit parade April 10, 1922. Fifty thousand citizens lined the route to bide farewell to the horses which had served the department since 1860. In this picture "Babe" and "Rusty" are pulling the hose wagon of Engine Company No. 37, at right. At left, "Pete," "Jim," and "Tom" are pulling an American LaFrance steamer. (Clarence C. Woodward photo.)

Apparatus of the Hamilton, Ontario Fire Department about 1900, with chief's buggy in foreground, ladder truck in top, rear.

Five-horse hitch of the Boston Fire Department responding to an alarm in winter days. The last of these units went out of service in the mid 1920s. (Boston Fire Department photo.)

Remembrance of Things Past [6]

From History Highlights of the Salt Lake City, Utah Fire Department by Retired Chief J. K. Piercey.

This information in regard to the days of the horse-drawn fire equipment comes from my memories of living approximately backyard to backyard of Number 4 Fire Station from the time of my birth. In my early youth I received many a free haircut at the station, and even though the horse clippers were used it was still quite presentable.

It was always a fascinating sight to see the horses perform their hitches. Three separate hitches were made at 7:00 p.m. each night. The horses returned to their stalls to make the second hitch, but remained under the harness after the second hitch while straw bedding was placed in their stalls and their evening oats and hay supplied.

The hitch consisted of the horses running underneath the harness, which was suspended from the ceiling on straps. As the firemen gave a downward pull on the collar, the hooks which held the harness to the straps would be released. Incidentally, the collar was open just opposite from the conventional collar. It opened underneath, and, as the horse came under, the firemen would snap it down over the neck.

After they were in the stalls, a special type of clay was packed into the sole of each hoof. The purpose of this was to prevent the hoofs from drying out. Due to the galloping of the horses it was necessary to keep their hoofs and legs in first class condition. This clay would readily break out of the hoofs upon the first few steps they took.

The horses in the Fire Service were outstanding specimens. They were intelligent and well trained. One could not help but note the feeling of pride of the firemen toward their horses.

The morning for the drivers started at 5:00 o'clock. They would be called by telephone, and between 5:00 and 5:50 a.m. they would clean the bedding from the stalls, curry their horses, get them

107

Typical of parades in days gone by was this one in Salt Lake City, Utah, in 1913. Small boy at left in costume of his time helps create a nostalgic scene repeated many, many times throughout North America. (Utah State Fire Marshal.)

their morning food and water, and the driver would be ready to go home for his breakfast at 5:50 a.m.

At 10:00 a.m. a hitch was performed and again at 12:00 noon, after which the horses were fed.

Although the gong sounded five times for a fire and seven times for a hitch, one could detect in the actions of the horses that they instinctively knew the time of the different hitches from an actual fire alarm.

The drivers returned from breakfast at 7:00 a.m. and at 8:00 they would take the horses out for a 45-minute exercise period. This consisted of walking the horses back and forth on Fifth Avenue

Engine No. 1 in Cleveland, Ohio pulling out of quarters. (Photo by Kim Carr.)

between "H" and "K" Streets, after which they were brought back and rubbed down.

If a fire alarm came in during the exercise period, one of the firemen would run to the corner and hail to the driver. It was quite a sight to see them galloping bareback to the station to be hitched and dash to the fire. The exercise period was held each day — Sundays excepted.

The floor of the fire station was made of wooden blocks. The stalls were kept scrupulously clean and showed evidence of being scrubbed with soap and water.

In the summer and fall, during the day, all blinds on the apparatus floor were drawn. This was a measure to keep the flies from annoying the horses.

After returning to the station from a fire the horses always received first consideration. They were washed down and dried and their hoofs carefully checked over before any attention was given to the men or equipment. The horses were always covered with blankets while standing at a fire. Of the hundreds of times that I have seen these fine animals dash out of the station and gallop down the street, the last time was just as thrilling a sight as was the first.

The ladder truck at the station, carrying ground ladders up to fifty feet, was pulled by two horses, and the hose wagon, containing 1,000 feet of 2½-inch hose and other small fire fighting equipment, was drawn by three horses.

October 15, 1917, saw the complete motorization of the Salt Lake City Fire Department and the remaining fire horses were retired to the city farm in Mountain Dell.

Street scene in Bismarck, North Dakota, at the turn of the century showing apparatus of Pioneer Fire Company No. 1, members of the department and local citizens. (D. E. Gilman, Secretary, NDFA.)

Cincinnati Incident[7]

An old-time visitor to Cincinnati furnished the following graphic description of the first steam fire engine in this city. "Yes, sir, I drove the team that hauled the first steam fire engine ever built to the first fire on which water streams were played by steam power. I'll tell you how it was.

"My brother worked in Miles Greenwood's foundry in Cincinnati and I lived at Island Pond, Vermont, and in May 1852, I went to Cincinnati to see him, arriving there Saturday evening. We were on our way to church Sunday morning when the fire bells struck, and my brother said, 'Now we'll see what they will do with the steam machine,' and we started for Miles Greenwood's shop where the first steam fire engine was. It was the first ever on wheels.

"There the engine stood, steam up, four large, gray horses hitched to it, and a crowd looking at it and Greenwood was mad as the devil because he couldn't get a man to drive the horses. You see, all the firemen were opposed to this new invention because they believed it would spoil their fun and nobody wanted to be stoned by them, and the horses were kicking about so that everybody was afraid on that account.

"My brother said, 'If you can, I wish you would.' So with my Sunday clothes on, I jumped on the back of the wheel horse, seized the rein, spoke to the horses and out we went kiting. Miles Greenwood went ahead telling the people to get out of the way because the streets were full of people.

The horses went on a fast run nearly all the way, and when we got to the fire, we took suction from the canal and played two streams on the building, a large frame house and put the fire out. This was the biggest crowd I ever saw in my life and the people yelled and shouted while the firemen who stood around the machine jeered and groaned. After the fire was out, Greenwood put on two more streams and the four of them were played. Then the City hired me to drive the four-horse team with the steamer paying me $75 a month. It was a great, long, wide affair with a tall, heavy boiler, and it ran on three wheels, two behind and one in front to guide it. After a few weeks, a fellow offered to do my work for $50 a month, and they turned me off and hired him. The second fire he drove to he was run over and killed."

Early 20th century fire scene in downtown Boston showing horses standing patiently near three steamers, a ladder truck and a water tower.

An 1860 fire engine built by the firm of Reaney and Neafie. (Smithsonian Institution Photo.)

A Silsby rotary steam fire pump of the 1860s. (Smithsonian Institution Photo.)

Steam Fire Engines[1]

Quite a few manufacturers invested in the production of steam fire engines. At least eighteen firms were building such apparatus by 1859. Some made the self-propelled type; others made engines to be drawn by two, three, even six horses; and some engines still were being designed to be pulled by manpower.

The Silsby Manufacturing Company, the Amoskeag Manufacturing Company, and the Clapp and Jones Company each designed a unique type of steam-powered pump that was to last until the next major change in apparatus design, toward the end of the century.

The Latta Company in Cincinnati, Reanie and Neafie in Philadelphia, the Portland Company of Portland, Maine, William Jeffers of Pawtucket, Rhode Island, and the Button Company of Waterford, New York also were producing steam fire engines in this period.

The Silsby Company started production in Seneca Falls, New York, in 1856. It made a rotary steamer, using a gear-type pump designed by Birdsall Holly. Despite a great deal of controversy over the merits of rotary and piston pumps, this firm continued to produce rotary pump steamers until it merged with other manufacturers in 1891.

Perhaps the most famous of steam fire engines were built by the Amoskeag Manufacturing Company which began production in Manchester, New Hampshire in 1859. All its steamers featured outstanding craftsmanship and performance and more than 800 were purchased by fire departments throughout the country. The first eleven Amoskeag engines had two steam pistons driving a rotary pump, but after a year or so, the company switched to making single and double piston types.

This steamer of Charles Ahrens and Company, Cincinnati, Ohio used first and second class Latta boilers.

A Latta steam fire engine on three wheels in St. Louis, Missouri, 1855-60. It was drawn by four horses. (Smithsonian Institution Photo.)

The Philadelphia firm of Reaney and Neafie constructed this steam fire engine for the fire department in Madison, Indiana. The draw bar at right indicates that the pumper could be pulled by horses or manpower.

This steamer was built in 1868 by John Ives and Son Company, Baltimore, Maryland, for showing at the Paris Exposition of 1868. It cost $9,000, sold for $4,000 and remained in service until 1890.

Original members of the Detroit, Michigan Fire Department pose with the city's first Amoskeag steam-powered fire engine. It was placed in service October 4, 1860 as "Lafayette Steam Fire Company No. 1." (Clarence C. Woodard photo.)

In the early morning of September 8, 1934, the luxury liner *Morro Castle* was approaching New York Harbor on her return trip from Havana, Cuba. It carried 316 passengers and a crew of 232. It had been built in 1930 at a cost of $4,800,000 and, with a sister ship, the *Oriente,* was considered the finest, most luxurious vessel ever placed in coastwise service. Unfortunately, the ship had a great deal of combustible material, including plywood partitions, elaborate trim, and draperies. The hull and superstructure were of steel but wood was also used in the superstructure and passengers' quarters. Sometime before three o'clock, fire was discovered in a locker near the writing room on the port side of the second deck. A steward notified the night watchman and returned to attack the fire with an extinguisher but the fire spread rapidly through the writing room into the main lounge which was two decks high, and then spread fore and aft on the upper two decks. The ship was brought about in an attempt to reach the Jersey shore and other ships nearby sped to rescue passengers. They managed to save 166 persons and others were picked from the water by smaller boats, but 124 persons died in the fire or drowned. The *Morro Castle* was beached at Asbury Park, New Jersey, as thousands watched from shore. (International News photo.)

1900-1940

Chapter V

"Much of the loss of life and property by fire in the United States is undoubtedly preventable. The extent and nature of our fire losses bring the question of fire waste prominently forward in connection with conservation effort." President William Howard Taft, May 30, 1911.[1]

Even though the 20th century began with the promise of scientific and technical achievement there was no diminishment in the vulnerability of American buildings to fire. From the early nineteen hundreds until the end of the 1930's, the number of fires and explosions that occurred, and their diversity, provided convincing evidence that the fire problem in America was unique, devastating, and in urgent need of correction.

In 1901 a conflagration wiped out 1,700 buildings in Jacksonville, Florida. A year later, in Paterson, New Jersey, 525 buildings were destroyed in one sweeping blaze. In December 1903, came the tragic Iroquois Theatre fire in Chicago which killed 602 persons. (See page 116).

Then came a conflagration in Baltimore, Maryland in which the lack of standardized fire hose coupling screw threads was pinpointed as an important reason for inadequate fire control measures. Two years later the famous earthquake in San Francisco occurred, providing another sound reason for practical building codes, enforcement of fire safety regulations and upgrading of municipal water supply. These were not the last of the conflagrations; such major fires would continue to occur for the next forty years, but their frequency seemed to diminish as fire resistive construction became more prevalent. But, as the threat of such sweeping fires seemed to lessen, the fire problems in individual occupancies became more important.

In the early part of the century, theaters were one of the principal locales of public entertainment.

Stage performances and the new motion pictures (without sound) brought large attendance, but even though terrible fires had occurred in such buildings in previous years, most of these structures had not been brought up to the known requirements of fire safety of that period. Consequently, they were likely to be buildings of tragedy — and tragic incidents did occur!

Another vulnerable group of occupancies included hospitals and institutions, each housing hundreds of helpless persons who were ill-equipped to respond to the sudden shock of a fire emergency. To some extent, these buildings were similar to theaters, in that their exit facilities, interior finish, and other elements of basic fire protection had not been updated.

Those early years of the 20th century also brought some tragic school fires and a number of explosions that resulted in extensive loss of life. Ironically, as these disasters were occurring, fire protection as a profession was advancing on every level, with developments in motorized fire apparatus, fire extinguishing agents, automatic equipment, communications, professional fire departments, fire prevention programs and public education.

This chapter includes a number of articles, some reprinted in much of their original form, to reflect the kind of fire analysis of those years and the intensity of reaction to needless fire tragedies. Superior figures refer to sources of these articles in the Bibliography on page 241.

The boat as it was beached off Hell Gate after 1,030 lives were lost. (Wide World photo.)

A large amount of combustibles, an untrained crew, bad storage of equipment and supplies, failure to practice lifeboat drills, and bad decisions by the ship's captain were the principal factors that contributed to this shocking fire tragedy. But today, three-quarters through the twentieth century, ships with one or more of these weaknesses are still being used as passenger vessels.

Tragic Trip

of the

"General Slocum"

At nine o'clock on the morning of June 15, 1904, the excursion steamer *General Slocum* with 1,400 men, women and children aboard steamed away from the pier at 13th Street and the East River in New York City. Most of the passengers were members of the Sunday School of St. Mark's Lutheran Church, 328 East Sixth Street, going to a summer picnic.

Just about half an hour after the side-wheeled, wooden vessel got under way, fire was discovered in the forward deck. The tragic sequence of incidents which followed served as a classic example to all persons responsible for fire protection aboard ship.

The crew brought out a hose line only to have it

Picture of the *General Slocum* on a typical excursion before the tragic fire. (Photo by Walter E. Hallam.)

Bodies of victims were placed on shore by firemen and police. The tragedy stunned the nation, but served as a classic example of the need for fire protection, particularly aboard ship. (Wide World photo.)

rupture when the water was turned on. Passengers who reached for life preservers found them to be stuffed with useless material instead of buoyant cork. Canvas wrappings and straps of the preservers tore when they were handled. People jumped into lifeboats, but these were lowered improperly and overturned before they reached the water and a lot of the passengers drowned.

A strong breeze was blowing and within minutes flames swept aft through the dry timbers of the old boat. Many women and children died when their flimsy dresses were involved by fire. The ship's captain, William Van Schaick, turned the bow of the vessel to North Brother Island and tried to race the vessel at top speed despite the fact that tugs and other boats were trying to catch up to the steamer. By the time the *General Slocum* was beached, 1,030 persons had died either from the flames or by drowning. Most of these individuals came from the "Little Germany" area between Houston and 23rd Streets from the East River to Fourth Avenue in New York City, and the tragedy wiped out a major part of that community.

Captain Van Schaick later was charged with neglect in failing to hold lifeboat drills, inspect fire hose and examine life preservers. He was convicted in the U. S. District Court and given a ten-

year sentence in Sing Sing Prison. He was pardoned by President William Howard Taft after serving two and one-half years.

Like other major fire tragedies, the *General Slocum* disaster brought subsequent improvement in safety to the public. Prior to the tragedy, regulations existed which required such passenger vessels to hold standard evacuation drills and fire drills, to have fire protection and fire fighting equipment as well as a trained crew. But, the coroner's inquest and a Federal investigating commission appointed by President Theodore Roosevelt disclosed that the *Slocum*'s crew was untrained for the emergency, that the ship's standpipe hose was hardly used, and that poor storage of hay, paints, oil and other combustibles made the ship ripe for disaster.

The December 1904 issue of *Munsey's Magazine* reported that, after the *General Slocum* fire, 268 ships in New York Harbor were inspected. More than one-third of them had defective life preservers, one-quarter had defective hose, and less than half had as many feet of hose as was required by the law.

The *General Slocum* fire demonstrated the terrible potential of a ship fire when the fundamentals of fire safety are overlooked or neglected. The fundamentals are just as important today.

A flammable curtain, a useless fire extinguisher, an asbestos curtain that would not drop and two automatic ventilators that were not completely installed — all of these contributed to the greatest loss of life in a theatre fire in the United States. This article is followed by a very moving eyewitness report by actor Eddie Foy.

The Iroquois Theatre [2]

The greatest loss of life in a theatre fire incident in the United States occurred on December 30, 1903, in Chicago, Illinois, when flames swept the stage and auditorium of the Iroquois Theatre, with a toll of 602 dead and 250 injured.

The building itself was of fire-resistive construction and was considered the last word in modern design. Even judged by present day standards, it would in many features be considered satisfactory. In certain features of fire protection, however, it seems to have been woefully lacking. An asbestos curtain had been provided to protect the audience in event of fire on the stage, but due to an obstruction was prevented from properly functioning. Two automatic ventilators had been installed to ventilate the stage, but they were not finished when the fire occurred and were fastened shut. Exits were not properly marked. The gallery did not have a main exit of its own. Fire protection appliances apparently were lacking and no fire alarm box had been installed. The attendants had not been drilled for emergencies. Sprinklers were not provided over the stage.

The seating capacity of the auditorium was 1,625, but on the afternoon of the fire 1830 persons are reported to have been in the audience and 275 on the stage. Two hundred, therefore, were seated or standing in the aisles.

The fire occurred during the second act of the play. It is said to have been caused by hot particles of carbon from an open arc light coming in contact with a border drop. The light was in use as a spotlight for flooding the stage and was operated in conjunction with a swinging reflector. Several futile attempts are said to have been made to extinguish the blaze in its incipiency by beating it out and by the use of dry chemical extinguishers, but both methods were ridiculously inadequate.

The flames spread through the tinder-like fabric of the scenery with terrible rapidity and within a minute the fire was beyond control of anything but a well-directed hose stream. Meanwhile in the audience those on the opposite side of the curtain could see the blaze and the men fighting it, and they began to get frightened. Someone yelled "Fire" and the mad rush for the exits began.

Vain efforts were made to drop the asbestos curtain, which descended part way. It was apparently obstructed by the proscenium light board, which was hinged to the wall and had swung out under the north end of the curtain. Instead of being fastened in a rigid frame sliding in grooves at the side, the curtain hung loose. The exit from the stage at the back was opened by those escaping, and this caused a strong draft which bellied the slack of the curtain in a wide arc out into the auditorium, letting the smoke and flame through at its sides. A ventilator over the main auditorium was open, as well as the highest alley exit at the rear of the gallery, and the draft was naturally increased in that direction.

As a result of these conditions the full volume of flame, smoke and gases swept through the stage opening and bore directly on the gallery and, to a less extent, on the balcony. The auditorium was exceptionally wide and shallow, which brought the gallery and balcony nearer than usual to the stage. Moreover, the fire was almost of the nature of an explosion. The stage equipment was so highly flammable that the fire reached its greatest degree of intensity very quickly and within fifteen minutes from the start had spent its force, having consumed most of the combustible material on the stage side of the proscenium wall.

It would appear that the greatest loss of life was from suffocation by smoke and direct action of the flames. Many were trampled to death as a result of panic. While the crowding due to panic unquestionably prevented the escape of many persons, the spread of fire was so rapid and severe that this was probably the immediate cause of death in most cases. The struggles of theatre patrons jammed on stairways and landings outside of the auditorium were terminated by smoke and heat. So swift was the spread of the fire that many in the gallery, who perhaps hesitated before making their exit, had apparently not moved from their seats before they were overcome.

Comparatively few persons on the main floor lost their lives, due to the availability of the exits and the fact that the fire was drawn upward by the draft. Of the 602 dead approximately 70 percent were in the gallery; the remainder were in the balcony. In addition to those lives lost, more than 250 were injured. The majority of the bodies

(continued on page 117)

Rhode Island Historical Society

The Burning of the Gaspée

On the night of June 9, 1772, a group of Rhode Islanders rode out to burn the British revenue schooner Gaspée. (See page 17.) The place where this occurred, seven miles south of Providence, Rhode Island, is known as Gaspée Point. The incident angered the British who later burned Jamestown, Rhode Island, in December 10, 1775, plus a considerable part of Newport which also borders on Narragansett Bay.

A reward of 500 pounds was offered for the names of the culprits who burned the Gaspée but the British were never able to identify anyone responsible for the deed.

The British also had an armed brigantine which bore the name Gaspée. This was captured after the fall of St. John, New Brunswick, and Montreal, Quebec, on the thirteenth of November, 1775, by American troops under the command of Brigadier General Richard Montgomery. The young general was later killed during the attack on Quebec in 1776.

The Rhode Island Historical Society at 52 Power Street in Providence has an original painting of the illustration above. It was completed in 1892 by Artist Charles DeWolf Brownell.

This 1783 Yankee Pumper No. 5 came from New York City and later Catskill. It has square cylinders and the compression chamber is rectangular. (All photos by American Museum of Fire Fighting.)

"Ready and Willing" is the name given to this bucket carriage of 1792 vintage. It was built for the City of Brooklyn in 1792, later transferred to the Village of Jamaica before the advent of water mains.

"Old Skiver 1812" was presented to the Museum by the Newtown Exempt Volunteer Firemen's Association. It originally served at Church and Vesey Streets in New York City.

It began with the Newsham . . .

. . . America's first successful mobile fire engine. This was the important change in fire fighting that helped bring an end to the task of filling hand buckets and tossing water on the flames. The hand engine still required filling by buckets, but new hose and the first small-tip nozzles were sufficient to introduce a new capability in fire control, the ability to aim a stream of water from a distance and move the stream around on the flames.

Engines of this type were built by Richard Newsham of London, England, who described them in his own advertising as follows:

Richard Newsham of Cloth's Fair, London, Engineer, makes the most substantial and convenient engines for quenching fires which carry continual streams with great force. The largest engine will go through a passage about three foot wide in complete working order without taking off or putting on anything and may be worked with ten men in said passage. One man can quickly with ease move the largest size about in the compass it stands in and is to be played without rocking upon any uneven ground, with hands and feet or feet only, which cannot be paralleled by any other sort whatsoever. There is conveniency for twenty men to apply their full strength and yet reserve both ends of the cistern clear from encumbrance that others at the same time may be pouring in the water which drains through large copper strainers. The staves that are fixed through levers along the sides of the engine for the men to work by allow alternate motions with quick returns yet will not spring and lose time in release. As to the treadles on which men work with their feet, there is no method so powerful with a like velocity or quickness and more natural state for the men. The five large sizes go upon wheels, well boxed with brass, fitted to strong iron axles, and the other is to be carried like a chair.

The Newsham (page 13), and the engines on these pages are part of the outstanding collection of apparatus, equipment and memorabilia on display in the beautiful American Museum of Fire Fighting of Hudson, New York, sponsored and operated by the Firemen's Association of the State of New York.

"Jefferson No. 26" was a side-brake piano style engine of the Volunteer Fire Department in New York City, known as the "Blue Boys."

This beautifully designed De Graw hose carriage of 1865 vintage was originally owned by the M. T. Brennan Hose Company No. 60 in New York City.

"The Volunteer Engine," built by Agnew and Company of Philadelphia in 1846 is one of the most ornate and powerful hand engines in existence.

Built by Edward B. Leverich this beautiful and costly parade hose carriage was donated to the Museum by Weiner Hose Company of Kingston.

This Ahrens-Fox 1928 front-mounted six cylinder pumper was presented by the Westfield, New York Fire Department.

Presented by the Gowanda, New York Fire Department this LaFrance steam pumper was designed to be drawn by a two- or three-horse hitch.

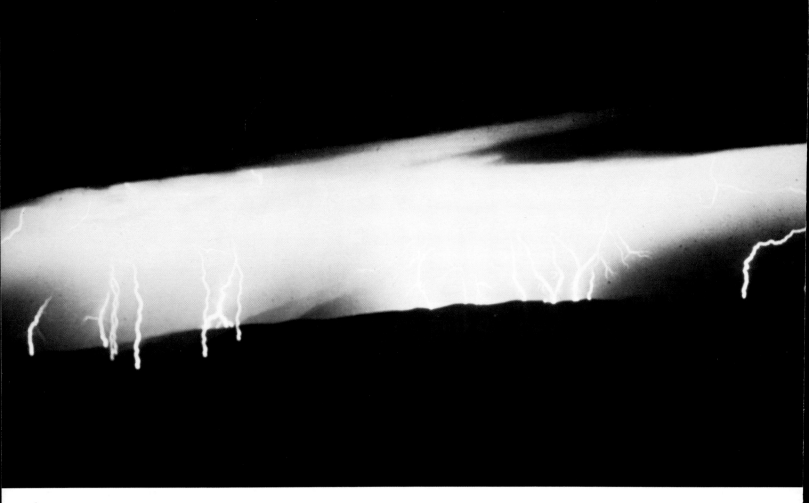

Multiple strokes of lightning are frequent ignition sources of outdoor fires. **(U. S. Forest Service photos.)**

Forest fires can develop intense flaming from ground to crown, moving rapidly on a broad front.

The Biggest Fires

Outdoor fires in forest, brush and grass areas usually are the largest and longest lasting of any that occur in North America. Many factors influence their size and intensity — type of fuel involved; slope of terrain; humidity; wind direction and velocity; accessibility to fire attack; mobilization; command and control of fire fighting forces; vehicles and fire suppressants available; the general logistics of food, equipment, basic supply, fuel, communications; and a host of other items.

Fires in extensive brush and forest lands may last for weeks, requiring mobilization of local, state and Federal fire control groups, establishment of central command, careful analysis of the fire's progress and potential, and skillful organization, planning and direction of the fire fighting crews. A large-scale "campaign" forest or brush fire is similar to a military operation in tactical and strategic use of available forces.

The largest concentration of fire fighting manpower and apparatus in history probably occurred in the State of California in 1970 when a long period of dry weather and heavy wind conditions caused hundreds of fires throughout the state.

There are about 180 fully paid and 700 volunteer fire departments in that state and practically every one of them was involved in mutual aid or direct fire control action in this dangerous emergency. Before the situation ended, more than 600 dwellings and other buildings were damaged or destroyed, and property loss exceeded one hundred and seventy-five million dollars. California is very susceptible to dry weather and high wind conditions, particularly in the southern portions, (see page 218) because extensive grasslands, brush and forested areas prevail the entire twelve hundred mile length of the state.

Lightning is a frequent cause of outdoor fires but its significance varies in the several states and provinces. In Florida about five percent of all recorded outdoor fires are caused by lightning. In Canada, the Department of Lands and Forests in Ontario attributed thirteen percent of outdoor fires to lightning. In Montana, Idaho, Washington, Oregon and British Columbia lightning strokes can be much more frequent and severe, sometimes starting thirty or forty simultaneous fires. When these fires occur in mountainous areas, or inaccessible wilderness, they may burn for days and weeks before control is established.

During the summer of 1967, the spectacular Sundance Fire in northern Idaho traveled sixteen miles in only nine hours, engulfing more than 50,000 acres of mountainous, timber-covered terrain. Spot fires were created ten to twelve miles beyond the origin and fire intensity built up to 22,500 British thermal units per second per foot of fire front, releasing nearly 500 million Btu/sec.

The fire created severe operational problems. The convection column rose to 35,000 feet and firebrands and debris lifted as high as 18,000 feet, then were transported by ninety-five mile per hour winds.

In areas where there is organized fire department protection, outdoor fires are generally handled by personnel responsible for structural fire fighting. In the more rural, or wilderness areas, a state forestry agency, the Forest Service of the U. S. Department of Agriculture, or the Bureau of Planned Management, U. S. Department of Interior may be responsible. The BLM protects twelve western states, including Alaska, a total land area of four hundred million acres. The Forest Service protects the national parks and remaining Federal properties throughout the states. State forestry agencies in the individual states protect

Practice . . .

. . . and the real thing!

Large aircraft are used for dropping fire retardants on key points of a spreading fire.

Helicopters are used for moving manpower and equipment and for dropping water and retardants.

five hundred million acres of woodlands under their jurisdiction.

Outdoor fires are best controlled by trained forces operating on the ground; but in the past twenty years, in the U. S. and Canada, aircraft have been used extensively to move manpower and equipment and to drop fire retardants to control fire spread. Helicopters and fixed-wing aircraft are active in a variety of operations, from simple reconnaissance to direct fire attack and extinguishment. "Smoke-jumpers" often drop by parachute to establish certain strategic fire control positions. Smokejumping has been an essential tactic since it was started in 1939. Aircraft have been used since 1915.

Short term and long term fire retardants, or fire suppressants, have specific applications, but they must be of a chemical nature that will not harm the ecology. The U. S. Forest Service and the Bureau of Land Management maintain continual research programs to improve means of fire protection, fire fighting and fire control. One of the programs in 1975 included evaluation of free-burning wild fires in 60,000 acres of wilderness land in Montana and Idaho to determine what would happen if these fires are allowed to burn freely. Similar experiments were underway in Louisiana and New Mexico.

USS Constellation

Some of the worst loss of life fire incidents have occurred aboard U. S. Navy vessels. On December 19, 1960, the *USS Constellation* aircraft carrier was being completed at the New York Naval Shipyard in Brooklyn when a forklift truck pushed a steel plate against the horizontal diesel fuel tank. The fuel spilled and ignited, involving the bomb shaft part of the hangar deck. In the next seventeen hours, fifty shipyard workers were fatally injured, 336 suffered non-fatal injuries, 40 New York City fire fighters were injured, and an estimated property loss of nearly forty million dollars resulted.

(Oil painting by Bollendock, Official U. S. Navy photograph.)

(World Wide photo.)

Here is the "Manhattan" one of the "Philadelphia" style handtubs manufactured in the first forty years of the nineteenth century. This style featured a central air chamber and brakes at the end with double "walking beams" or rocker arms that could be operated by additional manpower. (American LaFrance photo.)

This Button hand engine of the Red Jacket Veteran Firemen's Association, Cambridge, Massachusetts, became champion of the New England league in 1894. It was awarded the prize for being the handsomest in a tournament in Hartford, Connecticut, the following year. (American LaFrance photo.)

A crane-neck rotary steam fire engine built by the LaFrance Fire Engine Company of Elmira, New York. In an 1878 trial in Chicago, a third class engine of this type, starting from cold water, developed steam and fifteen pounds of pressure in two minutes; 105 psi in ten minutes. (American LaFrance photo.)

In Crescent City, Illinois, on June 21, 1970, at 6:20 a.m., sixteen cars of a moving freight train went off the tracks. Ten of twelve LP-Gas tank cars each carried 34,000 gallons of propane. Apparently the leading tank car was punctured as it left the tracks and the escaping gas resulted in this fire ball. Half of one tank car flew 800 feet through the air, shearing through a tree and two telephone poles, just missing fire fighters. More than 250 fire fighters and fifty-eight pieces of fire apparatus were called to handle the subsequent fires which destroyed twenty-three houses and sixteen businesses, and damaged eleven other dwellings. (Photo by Richard Anderson, Watseka, Illinois.)

On January 31, 1975, an American cargo ship, the Queeny crashed into the docked Liberian registered tank ship, the Corinthos. In seconds a major fire and series of explosions began, killing twenty-nine men, with subsequent injuries to twenty-five fire fighters and twenty-three civilians. The incident occurred on the Delaware River near Marcus Hook, Pennsylvania, and this scene shows the spectacular blaze that confronted responding fire fighters. (Photo by Thomas Kelly.)

On October 14, 1973, the City of Chelsea, Massachusetts, was again visited by a conflagration. On a warm day, with temperature at sixty-nine degrees F. and the northwest wind gusting up to forty-eight miles per hour, this fire quickly developed to a major conflagration very similar to its predecessor of April 12, 1908. The two incidents had essential features in common: immensity of fire area, fire load conditions, location of first ignition point, intensity of fire, violent air movement and fire brand spotting, and rapid spread during the main run or development. (Photo by Thomas J. Croke.)

THE AMERICAN PRIVATEER "GENERAL ARMSTRONG" CAPT. SAM. C. REID.

In the Harbor of Fayol (Azores) Octr 26th 1814. Repulsing the attack of 14 boats containing 400 men from the British Ships 'Plantagenet 74'_'Rota'44, and 'Carnation' 18 Guns. The 'General Armstrong' was 246 tons burthen Carried 6 Nine pounders and a 'Long Tom'(42 pounder;) amid ships and a crew of 90 men. The British loss was 120 killed and 130 wounded _ Americans lost 2 killed and 7 wounded

Capturing Moments in History

Nathaniel Currier, born in Roxbury, Massachusetts in 1813, became an apprentice in the shop of William and John Pendleton in Boston, two of the foremost lithographers in the new country. Currier quickly applied his artistic skills in his new trade and by the time he was twenty-one years old he moved to New York City and established his own lithography firm. In 1835 he published two lithographs that established his reputation as a foremost artist. The first had the rather unwieldy title: "Ruins of Planter's Hotel, New Orleans Which fell at Two O'Clock on the Morning of Fifteenth of May, 1835." But the lithograph that commanded great attention was that of the "Ruins of the Merchants Exchange, Destroyed in the Famous 1835 Fire." This was published four days after that event of December 16 and 17, 1835 and local acclaim for Currier launched his career.

For seventeen years he applied his unique artistry to capturing major events in history in the form of sketches and colored engravings. Then in 1852 he hired James Merritt Ives as a bookkeeper in the firm. Ives was a shrewd and capable businessman and a formal partnership was established between the two in 1857.

The firm used many important artists to carry on the special presentations developed by Nathaniel Currier. Currier was an active volunteer fireman and illustrated many incidents of different fire departments and volunteer fire companies in the eastern states. Altogether, the firm published about 7,000 prints of important events. Seven hundred and thirty-eight of these were hand colored, and 218 were uncolored. The original illustrations have become collector's items of high value.

Authentic reconstruction of Eagle Engine Company No. 13 stationed in New York City in the nineteenth century. (All photos by The Home Insurance Company.)

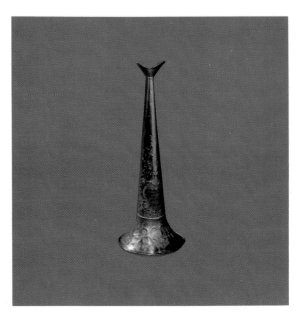

Ornate copper working trumpet used by William "Boss" Tweed a "Wheeler and dealer" leader of Tammany Hall. As a volunteer fire fighter he served as foreman of Americus Engine 6 of New York City.

Outstanding Collection

On the fifteenth floor of the office building at 59 Maiden Lane in New York City is one of the foremost collections of fire fighting equipment and related items. It belongs to the Fire Fighting Museum of the Home Insurance Company and was started as a hobby of the late Harold V. Smith who served as president of the company from 1937 to 1954. In addition to the items shown on this page, the Museum includes a large part of the collection formerly in the old museum of the Volunteer Firemen of New York.

There are more than 3,000 firemarks of which 1,431 different marks and variants represent fifty-six countries. The largest groups are from Great Britain and the United States, with designs cast in iron, lead, zinc, copper and brass. Walls of rooms are covered with helmets, ribbons, watchmen's signal rattles, badges, trumpets, helmet shields, and similar historic items.

There is also a display of valuable old prints depicting firemen in action, with wood cuts and engravings from the largest private collection in the world. Hand-drawn engines from the nineteenth century have been preserved in their authentic colors and furnishings. (See the Appendix for a list of addresses of this and other fire museums in the United States.)

Handtub with background of buckets and firemarks.

The handsome "Hope" engine, another of the "Philadelphia" style.

A beautifully maintained Amoskeag steamer, property of the fire department in The Dalles, Oregon. This was Number 529, last Amoskeag Harp Tank built by the Manchester Locomotive Works in New Hampshire. Purchased for $4,000 in 1879, this steamer and an 1879 Hunneman hand engine served as the only pumping units in The Dalles for seventy-five years. Last time the steamer was used was in the Hotel Albers blaze in 1918. Records indicate that 853 engines were built in the days of the Amoskeag-Locomotive Fire Engine Manufacturing Works in Hanover, New Hampshire. (Photo by Donald Sims.)

Hose reel and bucket carriage of the Richmond, Virginia Bureau of Fire which also has a 1906 Amoskeag, a 1922 Seagrave water tower and other antique apparatus and equipment on display. (Richmond Bureau of Fire photos.)

Loss of the Normandy

Just after the start of World War II, an almost unbelievable ship fire disaster occurred in New York City. On February 9, 1942, the *USS Lafayette* (formerly the French Liner *Normandie*) caught fire. In the next twelve hours fire fighting forces poured nearly a million gallons of water into the vessel in an attempt to halt the blaze. The fire was under control within the first five hours, but the ship was listing badly. In a little more than twelve hours, she capsized.

At the time of the fire, the *Lafayette* (or Normandie) was being converted from a passenger ship to a troop carrier. Even though she was one of the finest and best-known passenger vessels, a $60 million luxury liner, her construction fell short of safety standards in effect at that time. The U. S. Government took possession of the ship in May, 1941, and on February 9, 1942, a large force of workmen were engaged in refurbishing and equipping the vessel for service as a troop transport.

The fire started in the main salon amidships on the promenade deck when sparks from a welder's torch ignited piles of kapok life preservers. The disorganized and confused efforts at extinguishment which followed had "the elements of a Hollywood slapstick comedy" reported the NFPA *Quarterly* in July, 1942. Workmen attempted to fight the blaze and ten or fifteen minutes elapsed before the first alarm was received by the New York City Fire Department. Eventually, a 5-alarm response brought twenty-four pumpers, six ladder trucks, and three fireboats plus other miscellaneous equipment. Eventually, however, the "graceful old lady" just rolled over and died. She was later cut up for salvage.

In Cincinnati, Ohio, this fire drum was used at the start of the nineteenth century to sound fire alarms and other signals of major tragedy. It was succeeded by a watch tower, a bell (right) and finally a telegraph municipal alarm system. (All photos by Cincinnati Fire Department.)

A beautiful green patina has formed on this huge 2,000-pound bell that hung in a Cincinnati tower for many years. Made in 1866, it tolled alarms of fires and other signals until the start of the twentieth century.

Some years ago . . . you could sit in this chair at dusk . . . and listen to the sounds of the city closing down. . . . You might see the lamplighter reaching to start a flame in the gas mantle . . . you might hear four deep blasts from the whistle of a train . . . or sounds of metal wheels grinding over cobblestones. . . . There was time, then, . . . to think . . . to reminisce . . . to use the spitoon, if need be. . . . Just a quietude . . . while you wondered what the next fire alarm would bring . . .

Ten Factors That Contributed to the Iroquois Disaster

1. Absence of automatic sprinklers above the stage, required by a Chicago ordinance which was not enforced in any of the theaters.

2. Absence of effective first aid fire fighting equipment.

3. Obstruction of the asbestos curtain; it never got down within 12 feet of the stage for one or more reasons, and was later shown to have been of poor quality.

4. Blocking of exits with draperies, wood and glass doors, and barred iron shutters.

5. Complete lack of exit signs and directional exit signs.

6. A faulty exit pattern that channeled many persons into a single exit.

7. One fire escape was blocked by flames from an exit door lower down.

8. Electrical failure plunged the theater in darkness when the disaster was at its height.

9. Building 140 more seats into the theater (1,742) than were originally approved (1,602).

10. Selling of hundreds of standing room admissions (at least 200) resulting in further overloading the exit capacity of the balconies. Each of these deficiencies was laid to lack of vigilance by city officials charged with enforcement of the fire laws.

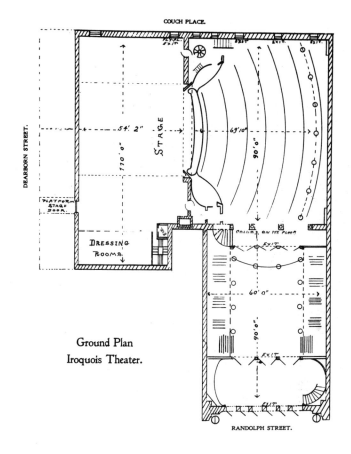

Ground Plan
Iroquois Theater.

were found in the aisles adjacent to the corridor from the balcony to the main entrance, piled three and four deep. The exits were not properly marked and when all the lights went out the audience could not see the exits to the fire escapes, and most of them sought to escape by the main entrance. The attendants had not been properly drilled and only two of ten alley exits at the north were opened. Each of these exits had glass panel doors and a pair of swinging iron shutters latched shut. The door at the front end of the top promenade leading to the stairway was locked.

The damage to the building was mainly confined to seats, scenery and furnishings. The fire destroyed everything on the stage that was combustible except some stage properties and the contents of a few dressing rooms that were cut off by a fire-resistive partition. Unprotected metal trusses above the stage were but slightly affected. This is accounted for by the fact that "flash" fire and air currents carried through the stage opening in the proscenium wall into the auditorium.

The superior character of the building construction apparently created overconfidence, which caused the absolute necessity for proper fire protection and ordinary precautions to be overlooked. The flammable contents of a fire-resistive building, without proper safeguards, are scarcely safer from destruction by fire than if they were in a building of ordinary construction.

117

The Iroquois — Eddie Foy's
Eyewitness Report[3]

Eddie Foy, one of the great personalities of the American theatre, was onstage when the Iroquois fire started. Here is his personal report.

After less than a year in The Wild Rose I opened on January 21, 1903, what was destined to be the most memorable engagement in my history. Klaw & Erlanger had recently taken up the producing of great extravaganzas — importing the big Drury Lane pantomimes from London and giving them an American farcical touch, a bit of additional lavishness. In splendor of staging and costumes they fairly out-Hendersoned Henderson. Their first production of this kind, The Sleeping Beauty and the Beast, had just run its course, and now they staged Mr. Bluebeard.

In this memorable production the Sister Anne of the original legend, who had been ruthlessly eliminated from Henderson's version, now appeared as the leading comedy character, and I was assigned to play it. I was supplied with two excellent songs by Billy Jerome, I'm a Poor, Unhappy Maid and Hamlet Was a Melancholy Dane, and I had a scene with a comic elephant which seemed to please the patrons greatly.

Newspapers exhausted their stock of adjectives in trying to describe this colossal affair. "Stupendous!" "Magnificent!" "The limit of pictorial stage art has been reached." "Extravagance can go no further." "One of the most gorgeous spectacles ever seen in a theatre" — such were a few of the comments.

There was a flying ballet which floated in the air like thistledown. At one point the leader soared from the stage out over the heads of the audience almost to the gallery rail, scattering flowers as she went.

The trolley wire upon which she was carried was destined to play a tragic part in the history of the show before the year was out.

At another moment in the play a girl on the stage stepped toward the footlights and held out her hand; one of the fairies floated down as gently as a snowflake, alighted thereon and pirouetted without causing the hand to shake.

There were some really wonderful tableaux: The Valley of Ferns; Egypt; India; Japan; The Parisian Rose Garden; and, The Triumph of the Fan. The last-named was the most gorgeous of all: hundreds of people on the stage, many in hand-painted costumes and waving scores of huge fans of ostrich plumes; greater fans of lace and feathers appearing from flies and wings, all illuminated with colored electric bulbs; and above all the seven aerial dancers, each poised on the points of an illuminated glass star which revolved under the touch of her toes.

Mr. Bluebeard opened the new Iroquois Theatre in Chicago on November 23, 1903.

In a Fool's Paradise

The theatre was one of the finest that had yet been built in this country — a palace of marble and plate glass, plush and mahogany and gilding. It had a magnificent promenade foyer, like an Old World palace hall, with a ceiling sixty feet from the floor and grand staircases ascending on either side. Backstage it was far and away the most commodious I had ever seen.

The space in the rear allowed for enormous expansion of the stage setting. They must have had as much room back there as they have in the Metropolitan Opera House in New York — perhaps more. A vast expanse of dressing-rooms was provided under the stage and auditorium for the chorus, while the principals dressed on the stage level or above. The flies were reached by elevators.

We were told that the theatre was the very last word in efficiency, convenience and, most important of all, in safety. It is true that the building itself was probably as nearly fireproof as a building can be made; but because of certain omissions — some careless and made in the interest of economy — it was a fool's paradise. There had been no great theatre disaster in this country for many years, and all precautions against such a thing were greatly relaxed.

Gorgeous But Dangerous

We drew big crowds all through Christmas week. On Wednesday afternoon, December 30th, at the bargain-price matinee, the house was packed, and many were standing. I tried to get passes for my wife and youngsters, but failed.

It was then decided that I should take only the eldest boy, Bryan, aged six, to the show and stow him wherever I could.

I made one final effort to get a seat for him down in front, but found that there were none left, so I put him on a little stool in the first entrance at the right of the stage — a sort of alcove near the switchboard — and he liked that even better than being down in the seats.

It struck me as I looked out over the crowd during the first act that I had never before seen so many women and children in an audience. Even the gallery was full of mothers and children. There were several parties of girls in their teens.

Teachers and college and high school students on their vacations were there in great numbers.

The house seated a few more than 1,600. The managers declared afterward that they sold only a few more than a hundred standing-room tickets, which would bring the total attendance to something over 1,700. The testimony of others indicated that there were many more standees than admitted by the management, and it was widely believed that there were at least 2,100 in the house — some reports claimed 2,300.

And remember that back of the curtain, counting the members of the company, stage hands and so on, there were fully 400 more.

Much of the scenery used was of a very flimsy character. Hanging suspended by a forest of ropes above the stage and so close together that they were well-nigh touching each other, were no less than 280 drops, several of which were necessary to each set; all painted with oil colors, the great majority of them cut into delicate lacery, and some of them of sheer gauze.

There had been a fire among the fluffy properties used in the big fan scene during our engagement in Cleveland, but by a piece of luck it was quickly subdued, and I had been playing in theatres for so long without any trouble with fire that the incident didn't give me much of a scare.

It takes a disaster to make one cautious.

After our experience at the Iroquois, not one in ten of us actors (and I dare say other people would have been equally heedless) could remember whether we had ever seen any fire extinguishers, fire hose, axes or other apparatus back of the stage. Some testimony was given which seemed to indicate that precautions of this sort had been woefully inadequate.

The play went merrily through the first act. At the beginning of the second act a double octette — eight men and eight women — had a very pretty number called In the Pale Moonlight. The stage was flooded with bluish light while they sang and danced. It was then that the trouble began.

In spite of some slight conflict of opinion, there can be no doubt that one of the big lights high up at one side of the stage blew out its fuse. That was what had caused the Cleveland blaze, and it was well known to the electricians of the company that in order to obtain the desired lighting effects they were carrying too heavy a load of power on the wires. Anyhow, a bit of the gauzy drapery caught fire at the right of the stage, some twelve or fifteen feet above the floor.

I was to come on in a few minutes for my turn with the comic elephant, and I was in my dressing-room making up, as I wore a slightly different outfit in this scene. I heard a commotion outside, and my first idle thought was, "I wonder if they're fighting down there again" — for there had been a row a few days before among the supers and stage hands. But the noise swelled in volume, and suddenly I became frightened. I jerked my door open, and instantly I knew there was something deadly wrong. It could be nothing else but fire!

My first thought was for Bryan, and I ran downstairs and around into the wings. Probably not forty seconds had elapsed since I heard the first commotion — but already the terror was beginning.

When the blaze was first discovered, two stage hands tried to extinguish it. One of them, it is said, strove to beat it out with a stick or a piece of canvas or something else, but it was too far above his head. Then he or the other man got one of those fire extinguishers consisting of a small tin tube of powder and tried to throw the stuff on the flame, but it was ridiculously inadequate. Meanwhile in the audience those far around on the opposite side, and especially those near the stage, could see the blaze and the men fighting it, and they began to get frightened.

The flame spread through those tinder-like fabrics with terrible rapidity. If the drop first ignited could have been instantly separated from the others, the calamity might have been averted, but that was impossible. Within a minute the flame was beyond possibility of control by anything but a fire hose. Probably not even a big fire extinguisher could have stopped it by that time.

Why no attempt was made to use any such apparatus, or whether, indeed, it was in working order, I don't know. If the house force had ever had any fire drills, there was no evidence of it in their actions. The stage manager was absent at the moment, and several of the stage hands were in a saloon across the street. No one had even taken the trouble to see that a fire alarm box was located in or near the theatre, and a stage hand ran all the way to South Water Street to turn in the alarm.

As I ran around back of the rear drop I could hear the murmur of excitement growing in the audience. Somebody had of course yelled "Fire" — there is almost always a fool of that species in an audience and sometimes several of them — and there are always hundreds of poorly balanced people who go crazy the moment they hear the word. I ran around into the wings, shouting for Bryan. The lower borders on that side were all aflame, and the blaze was leaping up into the flies. On the stage those brave boys and girls, bless them, were still singing and doing their steps, though the girls' voices were beginning to falter a little.

Foy's Greatest Role

I found my boy in his place, though getting much frightened. I seized him and started toward the rear. But all those women and children out in front haunted me — the hundreds of little ones who would be helpless, trodden underfoot in a panic. I must — I must do what I could to save them!

I tossed Bryan into the arms of a stage hand, crying, "Take my boy out!" I paused a moment to watch him running toward the rear doors; then I turned and ran out on the stage, right through the ranks of the octette, still tremblingly doing their part, though the scenery was blazing over them; but as I reached the footlights one of the girls fainted and one of the men picked her up and carried her off.

I was a grotesque figure to come before an audience at so serious an occasion; tights and comic shoes, a short smock — a sort of abbreviated Mother Hubbard — and a wig with a ridiculous little pigtail curving upward from the back of my head.

The crowd was beginning to surge toward the doors and already showing signs of a stampede — those on the lower floor not so badly frightened as those in the more dangerous balcony and gallery. Up there they were falling into panic.

Oh, if only I possessed an overmastering personality and eloquence that could quiet them! If only I could do fifty things at once — why didn't the asbestos curtain come down? I began shouting at the top of my voice, "Don't get excited. There's no danger. Take it easy" — and to Dillea, the orchestra leader, "Play! Start an overture — anything! But play!" . . . Some of his musicians were fleeing, but a few, and especially a fat German violinist, stuck nobly. . . . "Take your time, folks. (Wonder if that man got out with Bryan?) No danger!" — and sidewise into the wings, "The asbestos curtain! For God's sake, don't anybody know how to lower this curtain? Go slow, people! You'll get out!"

I stood perfectly still, and when addressing the audience spoke slowly, knowing that these signs of self-possession have a calming effect on a crowd. Those on the lower floor heard me and seemed to be reassured a little, but up above, and especially in the gallery, I could see them surging, fighting, milling about in the flickering light, a horde of maniacs.

Down came the curtain slowly, two-thirds of the way — and stopped, one end higher than the other, caught on the wire on which the girl made her flight over the audience, and which had just been raised into position for her coming feat! Instead of being fastened in a rigid frame sliding in grooves at the side, the curtain hung loose, and the strong draft, coming through the back doors by which the troupe were fleeing, bellied the slack of the curtain in a wide arc out into the auditorium, letting the draft and flame through its sides. "Lower it! Cut the wire!" I yelled. "Don't be frightened, folks; go slow! (Oh, God, maybe that man didn't take Bryan out!) No danger! Play, Dillea!"

Below me Dillea was still swinging his baton, and that brave, fat little German was still fiddling alone and furiously, but no one could hear him now, for the roar of the flames was added to the roar of the mob.

"Laugh, Pagliacci!" wrote Amy Leslie of the scene. "What if they do burn? Play on, clown, sing, dance, jerk about your funny petticoats, though your own hair is growing gray under that comical wig and deep wrinkles of fright and misery line your painted face. No magnificently armored knight in flashing steel and waving plume could stand a more heroic picture in the soul's eye of the grateful than this horror-stricken, courageous comedian who with every trick of gay pretense and active command endeavored to calm that panic-crazed and fire-dipped mass of fighting bodies without minds."

Then came a cyclonic blast of fire from the stage out into the auditorium — possibly a great mass of scenery suddenly ignited and fanned by a stronger gust, though some insist that the gas tanks exploded — a flash and a roar as when a heap of loose powder is fired all at once. A huge billow of flame leaped out past me and over me and seemed to reach even to the balconies. Many of the spectators described it as an "explosion" or "a great ball of fire." A shower of blazing fragments fell over me and set my wig smoldering. A fringe on the edge of the curtain just above my head was burning, and as I glanced up the curtain itself was disintegrating. It was thin and not wire-re-enforced: another cheat!

Now the last of the musicians fled. I could do nothing more — might as well go too. But by this time the inferno behind me was so terrible that I wondered whether I could escape that way; perhaps it were better through the auditorium. I hesitated momentarily, but Bryan had gone out by the rear — if he had gone out at all — and I was irresistibly drawn to follow, that I might learn his fate more quickly.

He Stuck to the Last

I think I was the last man on the stage; I fairly had to grope my way through flame and smoke to reach the Dearborn Street stage door, which was still jammed with our people getting out. Some of those dressing under the stage had to break down doors or escape through coal chutes. The actors and stage employees nearly all escaped — saved by the failure of the asbestos curtain to come down, which let the bulk of the flame roll out into the auditorium and brought death to many in the audience.

The flying ballet went out as I did, rescued through the heroism of the elevator boy, who ran his car up through tips of flame to the scorching flies where they stood awaiting their turn, and brought them down. But one of them, Nellie Read, the premiere, was so badly burned that she died in a hospital a day or two later.

As I left the stage the last of the ropes holding up the drops burned through, and with them the whole loft collapsed with a terrifying crash, bringing down tons of burning material — and at that all the lights in the house went out and another great balloon of flame leaped out into the auditorium, licking even the ceiling and killing scores who had not yet succeeded in escaping from the gallery.

The horror in the auditorium was beyond all description. There were thirty exits, but few of them were marked by lights; some even had heavy portieres over the doors, and some of the doors were locked or fastened with levers which no one knew how to work.

When one balcony exit was opened those who surged out on the platform found that they could not descend the steps because flames were leaping from the exit below them. Some painters in a building across a narrow court threw a ladder over to the platform. A man started crawling over it. One end of it slipped off the landing, and he fell, crushed, on the stones below. The painters then succeeded in bridging the gap with a plank, and just twelve people crossed that narrow footpath to safety.

Eight Minutes of Horror

The twelfth was pursued by a tongue of flame which dashed against the wall of the opposite building — and no more escaped. The iron platform was crowded with women and children. Some died right there; others crawled over the railing and fell to the pavement. The iron railings were actually torn off some of the platforms.

But it was inside the house that the greatest loss of life occurred, especially on the stairways leading down from the second balcony. The struggle there must have been one of the most hideous things in the history of the human race.

The stairways were one long mass of bodies, and wherever turns or landings caused a worse jam they were piled seven or eight feet deep. Firemen and police confronted a sickening task in disentangling them. An occasional body still breathing faintly was drawn from the heaps, but most of these were terribly injured. The heel prints on the dead faces mutely testified to the cruel fact that human animals stricken by terror are as mad and ruthless as stampeding cattle. Many bodies had the clothes torn from them, and some had the flesh trodden from their bones.

Never elsewhere did a great fire disaster occur so quickly. It is said that from the start of the fire until all the audience had either escaped or been killed or were lying maimed in the halls and alleys, the time was just eight minutes. In that eight minutes more than five hundred lives went out.

The fire department arrived quickly after the alarm and extinguished the fire in the auditorium so promptly that no more than the plush upholstery was burned off the seats, the wooden parts remaining intact. But when a fire chief thrust his head through a side exit and shouted, "Is anybody alive in here?" not a sound was heard in reply. The few not dead were insensible or dying.

Within ten minutes from the beginning of the fire, bodies were being laid in rows on the sidewalks, and all the ambulances and dead wagons in the city could not keep up with the ghastly harvest. Within twenty-four hours Chicago knew that at least 587 were dead, and fully as many more injured. Subsequent deaths among the injured brought the list up to 602.

As I rushed out of the theatre, I could think of nothing but my boy. I became more and more frightened; as I neared the street I was certain he hadn't got out. But when I reached the sidewalk and looked around wildly, there he was with his faithful friend, just outside the door. I seized him in my arms and turned toward the hotel. At that moment I longed only to see my family all together and to thank God that we were all still alive.

It was a thinly clad mob which poured out of the stage doors into the snow. The temperature was around zero, and an icy gale was howling through the streets. Many of the actors and actresses had had no opportunity to get street clothes or wraps, and some of the chorus girls who were dressing at the time of the fire were almost nude. Kindly people furnished wraps for these whenever they could and took them into business houses near by for refuge.

My own outfit of tights and thin smock felt like nothing at all, and my teeth were chattering so from the cold and the horror of what I had been through that I could not speak.

A well-dressed man, a stranger to me, stopped me and said, "My friend, you'd better borrow my coat," throwing off his heavy overcoat as he did so and helping me to put it on. He then picked up Bryan and walked with me across the street; and there, at the corner of a drug store, hurrying toward the theatre, I saw my wife with the two youngest children.

She gave a scream at sight of me, and crying, "Oh, thank God! Thank God!" she threw herself into my arms — then seized Bryan and kissed him, then me again, transferring quantities of grease paint from my face to her own and then to her son's. She had had a vague premonition of disaster from the time that Bryan and I left the hotel that afternoon.

Her ears unconsciously alert for significant sounds, she had heard the first clang of a fire truck gong in the street — rushed to the window, saw the direction it was taking, and felt certain that the trouble was at the theatre. Quickly she put wraps on the children and started toward the Iroquois on foot.

We turned back toward the hotel, thankful yet oppressed by the horror of the calamity which we knew must have occurred. I returned the overcoat to my good Samaritan friend, but was so agitated that I forgot to ask his name or even to thank him adequately, I fear.

I had no sleep at all that night. Newspaper reporters were begging me for interviews, friends were calling me by telephone and wiring me, and I had to reassure my mother and sisters and my wife's two sisters, who were taking care of our Eastern home. I was too excited to sleep, anyhow, even if I had had opportunity. My nerves did not subside to normal pitch for weeks afterward.

Baltimore —

One of the most important operational lessons for the Fire Service occurred in this conflagration. Because of the scope and rapid spread of this fire, the Baltimore Fire Department very quickly called for mutual aid assistance from a variety of cities and states. Response was excellent, but operations became difficult because of the different sizes and shapes of fire hose couplings, some of which simply would not fit any hydrants. The fire departments also learned how quickly a conflagration could spread through combustible buildings in a downtown district which featured many close exposures.

simply a matter of unmatched threads[4]

In Baltimore, Maryland, at 10:48 in the morning of Sunday, February 7, 1904, someone discovered fire in the basement of the John E. Hurst & Company building that stood on the south side of German (now Redwood) Street between Hopkins Place and Liberty. An alarm went immediately from the automatic box attached to the outside wall of the building. This brought response from Baltimore's Engine 15, a steam pumper and hose wagon (both pulled by two-horse hitches), Truck Company 2, the Fire Insurance Patrol and District Chief Levin H. Burkhardt.

Responding firemen started attacking the basement fire, but within seven minutes it spread rapidly up through an unenclosed well-hole in the 6-story brick building, bursting explosively from the top floor and involving nearby buildings which had unprotected window openings. For the next thirty hours, this conflagration burned completely out of control, destroying 155 acres (eighty city blocks), and 2,500 buildings, putting 50,000 people out of work, and causing an estimated financial loss of fifty million dollars.

George W. Horton was Chief Engineer of the fire department at the time, but within one-half hour after the fire started, he suffered an electric shock from a fallen trolley wire and command of the fire department fell to District Engineer August Emrich. Chief Emrich, Baltimore's Mayor Robert M. McLane, Pinkney W. Wilkinson, Secretary of the Board of Fire Commissioners and other city officials established a field headquarters building in the Salvage Corps Station.

It is well known that the Baltimore conflagration focused national attention on the need for standardized fire hose couplings and screw threads. As the fire progressed through the city, desperate calls for assistance were sent to Washington, Philadelphia, New York and other municipalities. Each of these cities sent horse-drawn apparatus on railroad freight cars. Apparatus and manpower from Philadelphia and Wilmington, Delaware, arrived in about two hours after being called. The New York train made a record run in a little over four hours. Washington, which first heard about the fire at 11:40 a.m., sent two engine companies, Nos. 3 and 6, by a special Baltimore and Ohio train which arrived about 1:30 in the afternoon. (Running time of this train was reported as 38 minutes just two minutes slower than the record train run for this distance.) Later Washington sent Engines 2 and 8 and additional manpower which totaled seventy-five men including Chief Engineer William T. Belt. Other cities sending apparatus and manpower included Chester, York, Altoona, Harrisburg, and Phoenixville, Pennsylvania; Annapolis, Sparrows Point, Relay and St. Denis, Maryland; and Atlantic City, New Jersey. Total manpower at the fire included 1,700 firemen of which 400 were unattached volunteers.

When they first arrived, out-of-town fire companies were delayed in attacking the fire partly because of the general confusion, and also because the many sizes and shapes of fire hose couplings just would not fit the hydrants. About forty fire companies went to the dock area and were able to draft from the waterfront. Others used barrels and wooden horse troughs set against hydrant openings, then their pumpers could draft from the miniature reservoirs thus formed.

What generally stunned the fire departments and other persons who saw the fire was the rapid spread from building to building and intense heat development even within fire-resistive structures. Weather conditions were normal with low winter temperatures but a strong wind blowing from 20 to 30 mph. The fire started in the heart of the business section, but because of wind direction and general lack of exposure protection, it spread easily from building to building. Hose streams from fire apparatus were practically useless against the intense heat. An NFPA report issued a few months later had this statement:

In contradistinction to ordinary fires in individual buildings which usually spread vertically from floor to floor, this conflagration was essentially a horizontal fire as regards its attack and progress in each building. As a rule every story was ignited simultaneously through the exterior windows and the fire swept across the building and out at the opposite side. Under these circumstances the protection of floor openings will avail but little if the windows are unprotected.

In a desperate attempt to halt the violent spread of fire, city officials authorized the dynamiting of structures. Two fire officers were designated by the mayor to select buildings that were to be demolished by explosives. Local building contractors were selected to carry out the demolition. Explosive charges were placed inside of buildings at the base of supporting columns and detonated by an electric "hot box." These tactics, however, did not stop the spread of the conflagration since even the largest charges of explosives failed to crumble a building. The heat of the fire was estimated at 2,200 to 2,500 degrees Fahrenheit.

Typical of newspaper descriptions of this famous conflagration is the paragraph below, taken from

Leslie's Weekly for February 18, 1904.

The night was black with the smoke and red with the flames as far as the eye could see. The furious gale tossed millions of great flaming cinders into the air. The panorama changed rapidly. Suddenly a great office structure would become brilliant, the light glaring through the windows as though every electric bulb and every gas jet in the building had been lighted at once. Then the dense, billowing veils of black smoke would hide it for a minute. Shortly a crashing sound would rise clear and distinct above the clamor and din and roar that were everywhere; and great leaping flags of flame would burst through the veil of the smoke, and float exultantly, it seemed, from the very top of the vast kettle of fire. In a few minutes more the building would be dark, and you would know that only the crumbling skeleton of it remained. You had seen a "fireproof" building burned out in half an hour! Then a new flame in a new quarter would arrest your attention, or the terrific heat would drive you from your post.

The conflagration finally died to controllable size when it reached Jones Falls, a 50-foot wide canal, in eastern Baltimore. Here about forty pieces of apparatus finally made an effective stop.

One of the amazing features of this conflagration was that no one was killed, although forty firemen were injured. Much hose and minor pieces of fire equipment were damaged but only one piece of apparatus was lost, Engine 15's pumper.

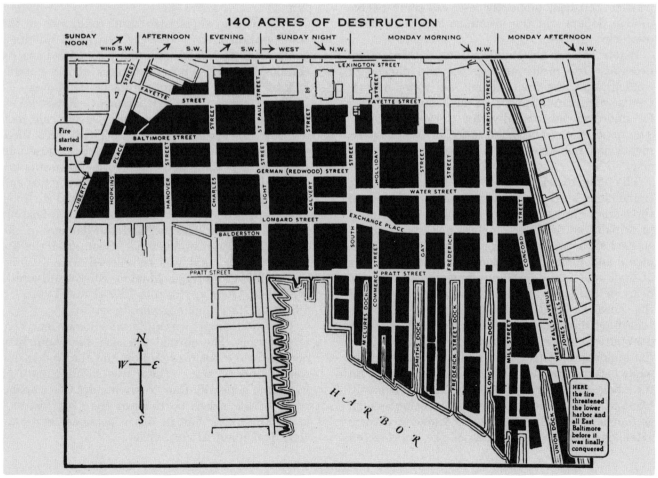

140 ACRES OF DESTRUCTION

Chelsea, Massachusetts, was devastated by a conflagration in 1908 because violations of common sense principles of fire protection were allowed by the municipal authorities. This was the first of two conflagrations that swept through Chelsea, Massachusetts, in the twentieth century. They were almost identical in fire start, fire spread and fire control problems. Each conflagration meant a serious financial loss to the city as well as destruction of a number of small business firms.

The Chelsea Conflagration [5]

On Sunday, April 12, 1908, a conflagration in Chelsea, Massachusetts destroyed approximately one-half the improved area of the city, about thirty-five hundred buildings, covering an area of nearly two hundred and seventy-five acres. Eighteen persons perished, monetary loss was some twelve million dollars and the insurance loss was a little over nine million.

Before the fire the municipal departments most necessary for protection against the danger of conflagration, the fire, building and water departments, were handicapped. Money actually needed for improvements was diverted to other channels. The maintenance and organization of these departments had not kept pace with the growth of the city.

The "rag district" where the fire obtained its tremendous start was covered with small one- to three-story frame buildings and shanties occupied principally as dwellings, stables and sheds, interspersed with two- and three-story rag shops, tenements and junk shops. Here and there long rows of brick and frame dwellings extended up into the heart of the city. Many of the buildings were of the cheapest and flimsiest construction. Approaching the business section, brick buildings predominated and on Broadway, the business center, the buildings averaged higher and the construction, while better than in the outlying districts, was, on the whole, rather poor. These buildings were mostly two and three stories in height, a few being as high as four stories. Sheds and other frame structures extended close up to the rear of the unprotected Broadway buildings. Brick churches, schools, city buildings and frame synagogues were located throughout the entire area, some with considerable open space about them. This space proved entirely ineffective in stopping the progress of the fire.

Viewed as a whole, structural conditions in the conflagration area were very poor, the buildings being mostly of wood with many concealed spaces, combustible roofs, boxed cornices and a notable lack of parapet walls and protected windows.

The Chelsea Fire Department consisted of twenty-one permanent and fifty-seven call men divided into seven companies. There were three steamers and one chemical engine with hose and the ordinary equipment. All the apparatus responded to a second alarm. Numerous cities and towns within a radius of ten miles could give aid in case of serious fire. There had been numerous fires in the "rag district" and the dangers of the locality were well understood. Soon after the fire was discovered, aid was summoned from Boston, Somerville, Cambridge, Malden, Melrose, Everett, Wakefield, Quincy, Newton, Revere and Lynn. In all, about thirty engines were in service.

The water supply was quite satisfactory for a city of this size. The normal pressure was about fifty pounds in the business section and the feed mains were of liberal size. There was a fair supply of hydrants, generally four to six hundred feet apart. Some of the newer portions of the "rag district" lacked hydrants but in other portions, hydrants were two hundred feet apart.

The ignition of rags and waste in the rear of Second Street near the corner of Carter Street started the fire. This section of the town at the extreme west was mostly open wet land in process of filling, but the usual dumps and numerous rag shops were in evidence. There were great piles of cotton and yarn waste, also woolen rags spread over the ground to dry. In some way unknown, these were ignited, causing a smoldering fire which had probably been burning some time. The wind, which was very high, was unusually severe in this locality owing to the open stretch of flat country extending over the marshes toward Everett. The burning rags were blown against a wax shop about two hundred feet distant, and set fire to it. Rags drying in the vacant space east of these buildings took fire and ten or fifteen minutes later the fire broke out in a large pile of rags inside a rag yard about nine hundred feet to the southeast of the Blacking factory.

The wind was very severe on this day; were it not for this the fire undoubtedly would have been controlled in its early stages. The maximum velocity of the wind at 9 o'clock was twenty-six miles an hour as recorded by the United States Weather Bureau at Boston. This increased to thirty-six miles as a maximum at 1 o'clock. It is possible that on account of the configuration of the land on each side of Chelsea the severity may have been somewhat greater here, for a wind of twenty-six to thirty-six miles an hour is not extraordinary for this section. These conditions, together with the poor construction and hazardous occupancy of this section of the city were the principal causes for the fire getting beyond control.

The fire started sometime after 10:30 and the first alarm was sounded at 10:45 a.m. The fire department on arriving proceeded leisurely to put a stream on the waste which was burning slowly, for the fire was then not at all serious. In about five minutes they noticed that the side of the wax building was also on fire and a second alarm was sounded at 10:55. This was more as a precautionary measure because of the high wind and the knowledge of the large amount of naphtha and rosin stored inside the yard. The chief went at once and in about ten minutes had the fire well in hand. Someone then came to him with the story that the rag yard on the corner of Maple and Elm Streets, about nine hundred feet away was on fire. He dispatched his chemical engine to this new fire. On arriving, the chemical man saw a brisk blaze in these rags, which were heaped up beside the rag shop. He was making progress and just considered that the blaze was practically under control when, to his surprise, the fire flashed through the office inside the main building. Understanding the seriousness of this,

General view of the conflagration from the northwest, about 11:30 a.m.

he sent word to the chief that this building was on fire inside and without waiting to investigate, the chief sent a special call to Boston for two steamers. This was received at 11:25. Then rapidly followed telephone calls to Boston and neighboring towns for all the help possible. About twenty-five engines responded up to 1 o'clock.

Meanwhile, driven by the high wind, the fire rapidly spread through two rag shops, both three-story frame structures, and by the time the first Boston engine arrived, possibly 11:30, was well on its way to Arlington Street. It swept through the small frame buildings in these blocks with extraordinary rapidity passing onward by leaps and bounds, picking out (wood) shingle roofs and obtaining a foothold on porches and buildings far ahead. The panic of the inhabitants who sought to remove their furniture and bedding to places of safety also helped in no small degree the spread of the fire. Many loads of furniture took fire in the streets as did that stored on sidewalks and parks.

View across Broadway Square about 4 p.m. during fire.

The chief had requested the Boston engines to rally on Arlington Street where he hoped that several brick buildings would serve to check the flames. When the engines arrived, the fire had already passed the street in places and was making great strides toward Broadway. It seemed impossible to stay its progress and the engines no sooner got in front of it than their position became untenable. By 12 o'clock it was passing Broadway near the junction of Washington Avenue and here a partial rally was made but nothing was gained and an engine had to be abandoned. By 12:30 it had crossed Broadway in many places and frame houses and sheds even as far in advance as Shawmut and Congress Avenues were on fire. It leaped whole blocks, and was beyond control. Many buildings in the residential section burned without any effort being made to save them. The extreme easterly section was sparsely covered and it was hoped for this reason that the specially hazardous buildings on lower Marginal Street and Eastern Avenue could be saved. The shower of sparks, pieces of wood, parts of buildings and the contents, all blazing, were carried by the wind far ahead. They settled down on the manufacturing plants, oil plants and East Boston buildings setting fire wherever they struck. Six Boston engines and the fire boat fell back on the East Boston shore and succeeded in putting out many roof fires. An oil works (storage plant) in Chelsea caught about 1 o'clock. This plant contained kerosene oil, naphtha and machine oils; several explosions took place.

Another burned about 2 o'clock and another about 2:30. Soon one of the tanks belonging to the latter company exploded, throwing a shower of burning oil over to the East Boston side of the trolley and railroad bridges and the Metropolitan Sewerage Pumping Station. These immediately took fire.

In the meantime, a creosote oil barge near the ship yard and the paint factory became ignited. The burning oil passed under the wharf of the latter and ignited it but the factory already had caught at the front, about 2 o'clock, from flying embers. Chelsea Creek was now covered with burning oil which was carried by an incoming tide up to the bridges and over to the wharves of the oil company on the East Boston side. The fire boat was hemmed in by burning oil but passed through it by directing its stream on the oil and making a passage for itself. The main buildings of the oil company caught from its wharves about 6 p.m. and were totally destroyed.

Thus in seven hours a stretch of land a mile and a half long and half a mile wide had been completely burned over.

On the whole, the fire seems to have been fairly well fought especially in certain sections. It is not strange that under the stress and excitement of a large conflagration many mistakes were made. If the entire energies of the department had been transferred to the first rag shop after it was discovered to be on fire the results might have been different, for the Blacking factory was so isolated that the fire there could hardly have menaced the city as a whole. A better trained and better managed department would have undoubtedly prevented much damage and might have saved the city in the early stages of the fire; but no department could have prevented the fire spreading to the waterfront after it attained proportions of a conflagration.

Conclusions (*From the original report*)

The most notable facts which this fire emphasizes are as follows:

1. The dangerous nature of pitch or mansard shingle roofs, frame porches, piazzas and accessory woodwork in spreading a conflagration.

2. The complete failure of any roof supported by unprotected steel or iron to withstand any but the smallest fire.

3. The need of good window protection where the sweep of the flames is parallel to division walls and the necessity of blank walls or properly protected window openings and parapet walls at right angles to prevailing winds.

4. The vulnerability of any ordinary buildings to sparks and embers, provided the bombardment be long enough, even though the space separating them from the burning buildings is great.

5. The slight value of streets of ordinary width in holding a fire when there is strong wind blowing and the fighting force is scattered.

6. The safest way to store oil in large quantities is in well made boiler iron riveted tanks having covers of the same material with large automatic relief valve, all well supported on brick or concrete piers.

7. Municipalities cannot violate the laws of good construction and fire protection without inviting conflagration.

8. The metropolitan water works system is shown to be exceedingly valuable for cities which it serves as it successfully withstood the extraordinary draught caused by this conflagration, although the Chelsea mains were not adequate in size nor properly gridironed.

9. More co-operation is needed between city officials and insurance interests in regard to protection against fire.

Chelsea cannot be considered blameless for this conflagration. The officials fully realized the conditions. Both water board and fire department had asked for improvements but the Aldermen refused to grant appropriations. Fire protection that is originally ample should keep pace with changed conditions in cities and almost invariably cities fail to recognize these changed conditions. In the case of Chelsea, however, it proved to be not so much defective water works and fire department as inadequate building laws poorly enforced.

Rag pile where fire started, in foreground Boston Blacking Company buildings at left, Hecla Compress Gas Company on right. Looking east.

General view of Bellingham Hill showing ruins of Highland School.

TORONTO FIRE RUINS, April 1904.

Toronto —

The City of Toronto, Ontario was victim of a conflagration on April 19 and 20, 1904. Starting in a neckwear manufacturing building in the downtown district, it swept through fourteen acres of the city, involving 123 business buildings and causing $12 million loss. Even though streets were 66 and 78 feet wide, the fire spread easily across these open spaces to involve other buildings. Six thousand people lost their jobs because of this devastation.

Theatre Panic — The Glen Cinema[6]

Another shocking lesson of the terrible results of panic occurred in the Glen Motion Picture Theatre in Paisley, Scotland, December 31, 1929. This was the winter holiday season and the theatre was overcrowded, exits were inadequate, not enough attendants were on duty and an iron grill covered the rear exit to the street. The theatre was the oldest in Paisley and had a seating capacity of about 1,100 adults although about 1,500 were attending the children's matinee on the afternoon of the disaster. The main floor was level without a pitch, and the balcony extending around three walls had two stairways to the main floor at the rear and two other stairways in front, leading to the stage.

The assistant film operator, a boy fifteen years old, was unwinding films in the spool room near the entrance of the theatre. He heard a hissing sound and discovered that film was burning in a tin box. He tried to carry it out of the theatre but dropped it near a door that was not used as an exit. He then ran to find the manager but a wind draft from the street swept the smoke into the theatre and people began to scream "Fire!" The manager, finding the box, opened an exit door and threw it into a vacant lot. By this time however, panic had seized the audience. Children in the balcony rushed to the stairway but finding descent impossible because of children crowding from the main floor, they leaped from the stairs and balcony on to the heads of the children below. Next they found that the double doorway leading to the outside was clogged by the struggling throng and that the outside doors were protected by the iron grill gate which was closed. They then ran to several anterooms and began to smash windows.

Altogether seventy children lost their lives and thirty-seven were injured sufficiently to require hospitalization. The death, in every instance, was due to suffocation from crushing and injuries received from trampling. Some of the bodies were lying three and four deep in the death trap corridor in the back of the stage.

It is hard to believe that the genius of Thomas Edison would not consider the possibility that fire might destroy some of his favorite accomplishments. But celluloid material, particularly nitrocellulose film, made a major contribution to the intensity and scope of this fire disaster.

The Edison Phonograph Works [7]

The portion of the West Orange, New Jersey, plant of Thomas A. Edison, Inc., involved in the December 9, 1914, fire was that known as the Edison Phonograph Works and covered most of the block bounded by Valley Road, Lakeside Ave., Watchung Ave. and Alden Street, West Orange, N. J. It consisted of twenty-two buildings, ranging in size and importance from a small one-story frame lumber storage building to a five-story reinforced concrete building covering a ground area of nearly 40,000 square feet. Of these twenty-two buildings, the fire spread through fifteen, destroying the buildings of wood, corrugated iron and brick construction, together with their contents, and seriously damaging the buildings of reinforced concrete construction, the contents of these latter being also generally destroyed.

The plant was devoted to the manufacture of disc and cylinder phonographs, disc and cylinder records, the business dictating machine, kinetoscopes or motion picture machines for theatres and the home, picture films (positives), the kinetophone, and other Edison devices and inventions. A large part of the plant consisted of machine shops for producing and assembling the metal parts of these machines, and buffing, lacquering, japanning, nickel plating and polishing shops for finishing the same; also an extensive woodworking department for manufacturing cabinets, with staining, filling, varnishing and finishing. Nearly half of the plant was devoted to the manufacture, testing and storage of the Edison disc and cylinder phonograph records. The disc records were made from a "condensite" compound prepared on the premises. The cylinders were made from celluloid stock purchased outside. Motion picture film manufacturing consisted of printing positives on sensitized film from negatives (taken in the Edison Studios elsewhere), developing, perforating, assembling, inspecting and storage.

There was also a group of ten buildings known as the Edison Laboratories, partly of brick construction and partly of corrugated iron and frame. Although somewhat exposed, these buildings were saved from the fire.

Briefly summarized, the Edison Phonograph Works consisted of a large number of buildings of exceptionally extensive area. Part of the buildings were of reinforced concrete, but none had standard fire stops between adjoining buildings, at floor openings or at windows to resist the horizontal and vertical spread of fire. They contained a large amount of highly combustible material and hazardous processes were not segregated, notably work on motion picture films. Under these conditions it is not surprising that the fire assumed more the proportions of a conflagration than of an ordinary blaze, and swept unchecked through the two main groups and some detached buildings, the fire protection being inadequate.

Special consideration later was given to the effect of the exceptionally severe fire on the reinforced concrete construction, notably the evidence of fusion of the concrete at a few locations where the fire was known to be hottest.

This fire added another disaster to the credit of the nitrocellulose material used in this case as a base for motion picture films.

The essential moral to be drawn is that fires of this intensity and magnitude must be made impossible rather than to expect the ideal in building construction.

129

It is well-known that working conditions were deplorable in many buildings of the early 1900s. The incredible disregard of simple fire safety practices helped take the lives of unknown thousands of people. Consider this typical manufacturing building that had inadequate fire exits, bad fire escapes, oil soaked floors and a quantity of gasoline. Note also the comments concerning enforcement of the city ordinance.

Tenant Manufacturing Building[8]

Twenty-five girls and women lost their lives in the fire which practically destroyed an old four-story brick and joist constructed building in Newark, New Jersey, November 26, 1910. The main building was four stories high with a blind attic five feet high at peak; adjoining sections were one- and three-stories high. This old structure was fifty by one hundred and thirty-two feet in area; walls were of brick; interior construction wooden girders and beams supported by unprotected cast-iron columns. Floors were double, two inches thick, but had been in use since the erection of building and were worn thin. The roof was constructed of composition roofing on boards with a wooden boxed cornice. There was practically no finish on inside walls, but about thirty per cent of the ceiling was sheathed with wood or metal. The one superimposed stairway was enclosed in a double board partition with doors arranged to close automatically. There was one elevator partly enclosed in frame partitions with glass windows; this, however, was of such flammable construction that elevator was practically open. There were also numerous small belt holes from floor to floor.

The building was heated by steam and lighted with electricity and gas. Steam power was derived from a one-story boiler house adjoining the main building. This boiler house was fairly cut off from the first story of main building. There was no watchman and practically no private protection. There was a good supply of fire pails at last inspection, but fire spread so rapidly as to make such protection entirely inadequate.

This building was equipped with two unreliable and inadequate fire escapes, one on the front of the building and one on the rear. There was an iron balcony covering two windows on each floor with steps inside of window from floor to window sill. These balconies were connected with straight up and down iron ladders so that only one person could start down at a time and these ladders were further located right in front of windows out of which flames and smoke were pouring when their use was greatly needed. During the start of fire one of these ladders became red hot, a veritable gridiron, the use of which was impossible, while the bottom ladder of the other fire escape failed to work through the sticking of a joint or ladder not being properly hooked on so that it could not be lowered and was practically worth'ess as a means of egress. In the rear, even had escape been possible, there was only a narrow alley difficult of access at such a crisis.

The occupancies of the building and those employed were as follows: First floor occupied for the manufacture of paper boxes, employing about thirty people, and by a machine shop employing about sixteen people; second floor, manufacturing of paper boxes, employing about fifty people; third floor, assembling incandescent electric lamps, employing about seventeen people, fire starting on this floor; the fourth or top floor was occupied for manufacturing underwear, employing from seventy-three to one hundred and fifteen people. Most of the employees in building were girls and women, there being only a few men employees in the building outside of the machine shop.

Fire occurred in this building on Saturday, November 26. Starting in the third floor lamp factory, it spread to gasolene and was immediately beyond control.

There appears to be little doubt that the fire was caused by upsetting a can of gasolene belonging to an apparatus for re-carbonizing the filaments of incandescent lamps. The filament to be re-carbonized was looped and connected with two poles supported in a vulanized cork or stopper which was placed in the mouth of a small metal can. A small iron pipe connected this can with the can of gasolene on the floor near the workbench, and the opposite side of the can on the table was connected with a vacuum pump. The air was first exhausted from the carbonizer, then the vapor from the gasolene was allowed to enter the carbonizer, and finally an electric current was sent through the filament in the carbonizer.

How the gasolene ignited the first time is not definitely known. A girl's shoe may have made a spark by striking a nail in the floor. A small electric motor was running in the room at the time, variously said to have been from twenty to fifty feet from the gasolene.

The main supply of gasolene for this process was kept in a barrel in the factory yard, and a supply was brought in several times a day in a gallon can. The supply can for the apparatus held about two quarts. This can had just been refilled by the girl in charge. When she saw the burning gasolene spreading over the floor she called for help. To get any sand a trip had to be made up the spiral stairway to floor above. That alone gave the flames ample time to spread. Two pails of sand were thrown on the blaze by an employee. The sand seemed to extinguish the fire, but apparently it only smothered it, for it suddenly flashed up a second time and quickly jumped to the ceiling.

A fire station was directly opposite the building. Instead of immediately giving the alarm, the foreman, the girls assisting him, tried to smother the flames with sand. With the rising flames spreading fiercely round them, they found it impossible to do what possibly might have been done a few minutes before when the fire was in the incipient stage. Finding their endeavors vain, a fireman who was standing outside the station was beckoned to, and he, not knowing the true state of the case, ran across with a chemical extinguisher. Too much time, however, had been lost and the flames had gained great headway. The fireman rushed back to the house of Engine 4 and turned in an alarm. The first alarm was received at 9:26 a.m., followed by a telephone call at 9:29 for a special truck from the fire house across the street from the fire. At 9:32 a second alarm was sent in and at 9:43 the third alarm was given. Although a fire engine company was directly opposite and was in its quarters at the time, the fire gained a great hold on the two upper stories before the first line of hose could be stretched. Before any water was thrown, the fire burned its way into the concealed spaces in the joisted floor, and each time it came through the sheathing it is said that there was a flash and a noise like an explosion.

The flames spread quickly through the building, at once creating a panic. The employees on the ground and second floors escaped without much trouble; some even on the third floor managed to reach the street in safety. With the hundred employees on the fourth floor it was very different. Choking and blinded by the smoke, they groped to find a means of exit. They tried the stairway but were driven back by the suffocating fumes of gasolene and smoke combined and many of the girls were forced to jump from the fourth story windows into the street. Others were overcome by smoke and died in the building. The fire spread so rapidly that the girls could not be prevented from jumping before arrival of the ladder trucks. Some rushed to the one miserable pretense of a fire escape near the end of the building, as some from the lower floors had escaped at that end, but the looped-up fire escape ladder had become too hot to handle and they had to be gotten down by means of a stepladder. At the rear of the building was the narrow, straight-up two-story fire escape, running down past the windows out of which the flames were issuing which had heated the ladder red hot.

Had the employees only stayed where they were they would have been rescued by the firemen, who by this time were raising the long ladders — a work of difficulty at the end of the building, on account of the narrow alley — and were extending the life nets to catch the girls if they jumped.

The firemen directed all of their efforts toward saving the women in the building and the fire was not checked until the building had been completely destroyed, with the exception of portions of the outside brick walls. It is reported that the dangerous condition of the building had been for some time widely known, although it had recently been passed as complying with the New Jersey State laws by Deputy Factory Inspector.

The building was built fifty years ago and was thoroughly saturated with oil. The flames spread very rapidly and cut off escape down the stairway from the fourth to the third floor before the employees on the fourth floor were aware of the danger.

The coroner's jury which investigated the holocaust, it is stated, was compelled to find that no one was to blame for the great loss of life even with the acknowledged deficiencies in fire escapes, overcrowding of the old fire trap and lack of facilities for escape of employees in case of fire. The allowing of the building, unfit for either factory or residential occupancies, to be occupied for the several hazardous occupancies employing the large number of girls and women — with the most hazardous occupancy on the third floor under the manufactory employing over one hundred women at time fire occurred — is grossly criminal.

The superseding of the city ordinance by the action of State Legislature which lowered the standard for protection of life and property and made such a verdict possible is not in accord with progress. Had either the city ordinance or state laws been intelligently enforced, however, the protection given would have eliminated the possibility of such a holocaust.

MISNOMER — The famous factory fire of March 25, 1911, has been called "The Triangle Shirtwaist Fire" through the years, and perhaps that was the easiest designation to remember. Actually, the Triangle Shirtwaist Company was only one concern on the 8th, 9th and 10th floors of the Asch Building on the corner of Greene and Washington Streets in New York City. The fire began in a rag bin on the eighth floor of a typical factory building of the era. Within thirty minutes about 150 young women and men burned to death or jumped to the sidewalks, a hundred feet below, or down elevator shafts. A few tried to leap into a 14-foot rope life net held by New York fire fighters, but the net broke and these victims died. At the time, New York City had no laws requiring fire drills, fire escapes, or sprinklers in factories. The owners of the Triangle Shirtwaist Company were indicted for manslaughter, but were found not guilty in a jury trial. The editorial below expressed the opinion of a national quarterly of that time.[9]

Fire Escapes[10]

The fire in the Asch building in New York City, and its attendant fearful loss of life, has precipitated a nation-wide discussion of the question of adequate methods of egress from buildings.

It has long been recognized that the common outside form of iron ladder-like stairway anchored to the side of the building is a pitiful delusion. This device for a quarter of a century has contributed the principal element of tragedy to all fires where panic resulted. Passing successively the window openings of each floor, tongues of flame issuing from the window of any one floor cut off the descent of all on floors above it. Iron is quickly heated and is a good conductor of heat, and expansion of the bolts, stays, and fastenings soon pulls the framework loose, so that the weight of a single body may precipitate it into the street or alley. Many a human being has grasped the hot rail of such a "fire escape" only to release it with a scream and leap from it in agony. Its platforms are usually pitifully small, and a rush to them from several floors at once jams and chokes them hopelessly. It is a makeshift creation of the cupidity of landlords, frequently rendered still more useless by the ignorance of tenants, who clutter it up with milk bottles, ice boxes and other obstructions.

Stair towers represent the only type of stair fire escape which cannot be rendered useless by fire inside the building it is designed to protect. If standpipes are placed in such stairways, with a hose equipment on each floor, they afford a safe and excellent vantage ground for fire fighting, as soon as the occupants of the building have passed out.

Though vastly superior to the common iron fire escape, even such egress as is afforded by these enclosed stairways may not always be adequate to prevent loss of life under panic conditions. Any stairway can be choked by panic-stricken human beings, so that a quick-burning fire will catch some of them who cannot crowd into the tower. The best safeguard is a fire wall running straight through every large area, dividing each floor into two fire sections, and having fire doors in the opening. Then when fire occurs in either section the occupants can quickly pass into the other, close the fire door, and proceed to the street in safety and without panic.

I have ten little toes
That are all in place;
I have one little nose
That is on my face.
My fingers are still
  Where they ought to be
And there's nothing at all
  The matter with me.

My thumbs and ears
  Are as good as new
So I'll give three cheers
  And a tiger, too!
I can dance and sing,
  I can shout with glee,
For there's not a thing
  The matter with me.
    – S.S. Kiser. –

A JULY 5th INVENTORY

Fireworks — A Dangerous Means of Celebration

Fireworks and human beings do not mix well. Thousands of victims of firework burns and injury can verify that statement. But each year when July 4 approaches, a kind of early summer madness begins all over the United States as adolescents and their older parents and friends hasten to buy or otherwise acquire fireworks to celebrate the anniversary of American Independence. On July 5 quite a few of these persons have changed from celebrants to patients in a medical clinic or hospital. Some die.

A reading of any summary of fireworks accidents is depressing, because the consequences of humans gambling with their personal safety shock our concepts of intelligent action. . . . A ten-year-old boy holding a large firecracker between his teeth when it explodes. . . . Hundreds of children with one or more eyes blinded by untimely explosions. . . . A man carrying a lapful of small bombs in his lap when someone playfully lights one firecracker. . . . The record of such incidents is an incredible encyclopedia of human folly.

In 1937, the year before a model ordinance for the control of fireworks was developed, an estimated 2,300 fires were caused by fireworks, with an unknown toll of injuries. In 1969, when a major national survey was made, there were 2,009 reported incidents, with injuries to 1,330 people and 774 cases of property damage. Five of the injured victims died.

Twenty-four percent of the injuries were caused by Class B fireworks which include toy torpedoes, railway torpedoes, certain firecrackers, and other devices. Forty-two percent of the injuries were caused by Class C fireworks which are not regulated by Federal legislation.

(The cartoon and verse above were published in 1916.)[11]

133

The tremendous power of a gas explosion is described and illustrated in this article. This particular gas, invented by Richard Pintsch, a German scientist, was a compressed gas obtained by cracking gas oil, consisting chiefly of hydrocarbon and hydrogen. At the time of this disaster such gas was used for lighting railroad cars and buoys.

Storage Battery House[12]

A disastrous explosion in New York City on December 19, 1910, killed ten persons and injured more than one hundred and twenty-five, wrecked the storage battery house in the Grand Central Terminal yard of the New York Central and Hudson River Railroad and caused considerable damage to adjacent buildings. The easterly part of the storage house was destroyed; a street car passing on Lexington Avenue was wrecked and overturned on a passing automobile truck; houses on Lexington Avenue and Fiftieth Street were partly shattered; and minor damage was done over an area within a quarter-mile radius of the battery house.

The storage battery house was two stories high, located on the southwest corner of Fiftieth Street and Lexington Avenue, extending along Fiftieth Street west from Lexington Avenue about one hundred and seventy-five feet, and was from thirty-five to forty feet deep to the south on Lexington Avenue. The ground floor was occupied as a carpenter shop; the upper floors as a storage battery room. The depressed yards of the New York Central extended to this building on the south and the area below the building, this area being entirely open on the south facing the yard.

The storage tracks in the depressed yard ran north and south to house and ended at ordinary timber bumper posts about at the south line of the house. While standing on these tracks the railway cars were charged with Pintsch gas. A gas storage and distributing system was located about five hundred feet to the south of building where the gas was received by tank cars from the Mott Haven yard and transferred to the storage tanks, service pipes leading from tanks over yard to charging connections. One of these pipes together with compressed air and steam pipes, ran parallel with the length of building in the open space under the battery house.

At about 8 a.m. a train of multiple unit electric cars in the terminal were being run down against bumper on a track about in line with the second row of columns from the east end of the battery house. The motorman failed to stop the train in time and it collided with and broke through the bumping post. Just back of the front columns in the rear of the bumping post ran the set of three or four pipes carrying the steam, compressed air and Pintsch gas. Debris or stored material struck by the train broke the near-by pipe carrying the Pintsch gas and during the next half hour a large column of gas escaped through the break.

The theory generally accepted as the cause of the explosion was that Pintsch gas escaping from a broken pipe filled the open spaces under the building with a mixture of the gas and air; that this gas and air mixture was ignited in some manner.

The explosion occurred in or about the battery house. The north wall of the house was thrown down into Fiftieth Street, the upper half of the east wall was blown down, and the floors and roof and roof-trusses fell, the latter hanging from the south wall, which remained largely intact.

The theory held by the authorities is that the gas accumulated in the space under the battery house and exploded there, although this space was open on its southerly long side and the gas was free to escape at this side. This theory does not clearly account for the violence of the explosion or the distribution of damage. Little damage was done immediately in front of the open basement space and practically no flame effects are reported from here. The gas storage station and a dynamite magazine south of it were not injured. Early rumors were to the effect that some dynamite was stored in the battery house but this was denied by one of the executive officers of the railway.

The damage done outside the wrecked building does not appear to be significant except in a few points. A two-story brick fire house, directly across Lexington Avenue from the wrecked battery house, had the parapet of its west wall, three or four feet high, broken and overthrown on a good part of the length of the building. Lower down, this wall had marks of being struck by several wooden beams, one of which in fact remained sticking in the brickwork. A street car passing between the battery house and the fire house was blown over and badly smashed. North of the east end of the battery house, a tenement on the northwest corner had marks of damage on the cornice of the south wall, and one of the top story windows lacked a triangle of brickwork over the opening. The store front on the ground floor of this house was wrecked. Practically all the damage beyond this consisted in breaking windows, shaking down plaster ceilings, and in a few cases forcing in window and door frames by pressure or shaking down arches of door openings. One building several blocks away from the battery house was found cracked. The area within which windows were broken extends much farther south than north, which may be significant in view of the fact that the space under the building in which the gas is supposed to have collected and exploded is open toward the south. The cars in the railway yard just south of the battery house were not damaged extensively, but their ends and not their sides were toward the explosive wave.

The *Engineering News*, in commenting on the force of the explosion and its effect, gives the following:

"The even distribution of the damage over the whole of the building affected is characteristic of a gas explosion. Where a charge of dynamite is set off, the destructive effect is greatest in the immediate neighborhood of the explosive and rapidly diminishes away from this center.

"Some incredulity has been expressed in the public prints as to a gas explosion being of sufficient force to do such extensive damage. It is little realized, apparently, what enormous potential energy is contained in even a moderate volume of combustible gas if mixed in proper proportions with air and exploded. A brief computation may be of interest, therefore.

"Suppose we take a volume of 100 cubic feet of gas of 600 Btu per cubic foot calorific power. When mixed with air in the proportions for most violent explosion, the mixture could be contained in a small room about twelve feet cube. The total heat units in this gas would be 60,000 Btu and the total foot-pounds of potential energy would be 60,000 by 772, equaling 46,320,000 foot-pounds. If this gas were exploded in small quantities behind a piston in a gas engine cylinder, something like fifteen to twenty-five per cent of this could actually be obtained as mechanical energy; and given proper mixture, something like this energy will actually be generated when the gas is exploded in a confined or partially confined space and will be exerted against any obstacle. This means that only 100 cubic feet of gas exploding when mixed with air may have a potential destructive power of over 10,000,000 foot-pounds; or it could lift a structure weighing a thousand tons to a height of five feet.

"Of course, under actual conditions in accidental gas explosions, the mixture of gas and air is imperfect, too rich in some places for greatest explosive efficiency and not rich enough in others. It will be apparent, however, from the above computation, that even a very few cubic feet of combustible gas mixed with air and fired has potential energy sufficient to cause an explosion of most destructive effect."

The surmise has been advanced that hydrogen evolved from the storage batteries accumulated in the battery room and exploded, but the best engineering opinion is that this theory is untenable.

View of wrecked battery house as seen from Lexington Avenue. (Engineering News photo.)

A number of important fire protection lessons were learned in the destruction of this supposedly "fireproof" building in New York City in which six people died. These are summarized at the end of this article. Fourteen years later, a new Equitable Building also suffered a fire loss (see page 151.)

The Equitable Building Fire[13]

The Equitable Building in New York City which burned on the morning of January 9, 1912, was a composite structure of five buildings erected at different times. It was of so-called fireproof construction, having floors mainly of wood on brick or hollow tile arches between wrought-iron and steel I-beams. The beams rested on walls and on columns mostly of cast-iron. The building was vitally weak in the fact that the structural metal work was not fireproofed and the cast-iron columns failed, causing three separate portions to collapse.

The first building of the Equitable Life Assurance Society was a five-story structure erected in 1869, with frontage on Broadway and Cedar Street. In 1886 the south wall was taken down and the building extended along Broadway to Pine Street. The height of the original building was increased to eight stories to correspond with the extension. An extension 50 feet wide was also built through to Nassau Street. In all this work the same general scheme of architecture was carried out except such improvements were made in construction as the progress of the building art would seem to warrant, notably the use of tile floor arches in place of brick. The height of the reconstructed building was equivalent to ten stories. The small lower buildings completing the block and located at the northeast and southeast corners were subsequently acquired by the Equitable Society and openings were made through the dividing walls, converting the whole block into one fire area. The entire block was approximately 48,000 square feet in area.

The greater part of the main building was gutted by the fire, which burned through all floors above the grade. A few areas such as the Cafe Savarin dining rooms on the grade and first floors, and the offices in the east end of the building up to the fourth floor, escaped the fire but were considerably damaged by water. With the above exceptions the destruction was practically complete as regards everything — combustible or noncombustible — in the nature of contents, trim and finish.

The structural work on the Cedar Street side above the third and fourth floors and on the Pine Street side above the sixth floor was badly damaged by the fire; at these points the beams and girders were generally deflected and likewise the columns are buckled or broken. The Cedar Street side shows the worst damage in this respect. The structural work in the east extension suffered little; there were a few deflected beams.

In three separate portions of the building the floors collapsed, apparently due to the failure of unprotected cast-iron columns. The largest collapse carried all floors and the roof down to the basement. In the middle section, the floors from the second to the eighth fell, leaving the roof standing. At part of the east collapse near the main stairway the roof and all floors above the third fell; in the remaining portion the collapse extended down to the second floor.

The seven-story building on Pine and Nassau Streets was burned over the upper two floors. Otherwise the building suffered only by water. Its escape was due partly to the fact that there were no communications with the main building on the lower two floors and the communications to the other unburned floors were indirect. The structural work of this building was uninjured.

The six-story building on Nassau and Cedar Streets communicated with the main building on all floors, yet the fire did not enter the former and practically the only damage done was by water. The work of the fire department was most effective at this end.

In the destruction of six lives and large property value by this fire the unprotected floor opening has added another to its list of horrors.

Aside from the striking evidence that a so-called fireproof building occupied chiefly for offices may foster a quickly spreading fire, most of the lessons pointed out in this case have been demonstrated by former calamities.

The supposedly favorable condition incident to the existence of high fireproof buildings across unusually narrow streets and each equipped with one or more standpipes also furnished a new and interesting experience in fire fighting which was not altogether satisfactory, having particular reference to the inadequacy in size of some of the standpipes.

The abnormally large amount of securities and valuables in vaults and safes exposed to the fire and impact as well as the weight of the building collapse was an unusual condition which attracted wide interest. None of the more important vaults were severely tested by heat. The safe deposit vaults were buried under the full weight of the principal floor collapse and resisted it well. None of the vaults was waterproof.

Although the fire department is entitled to the credit of having done all that could be expected with the facilities available, it is apparent that the fire, at least above the level of the fifth floor, burned unrestrained throughout the entire area of the main building. The high pressure service being too remote to be promptly available, it is difficult to figure how such a fire, if started again under similar circumstances, could be fought with any greater effectiveness by the fire department when the conditions of unprotected floor openings, combustible material, excessive and undivided floor areas, high wind, freezing weather, small standpipes in buildings across streets and the relative inefficiency of portable steam engines in comparison with the high pressure service are taken into consideration. If such a fire were started during business hours the work of the fire department would be impeded by the congested condition of the narrow streets and the endeavors of the firemen to rescue the numerous occupants whose only means of escape — a single stairway — might be cut off, as it was in the case of the three employees on the upper floors who lost their lives. The collapse of cast-iron columns was responsible for the loss of three additional lives.

This fire, like those in the Parker Building, Triangle Shirt Waist Factory and Alwyn Court Apartment House, calls attention to the inability of any fire department to fight a fire effectively which has once gained headway in the upper stories of a tall building lacking such essential fire appliances as an adequate standpipe equipment in conjunction with smoke-proof stair towers. The height of buildings should be limited in proportion to the effectiveness of their fire protection if life and property are to be conserved.

The timekeeper of the Cafe Savarin arrived at the building about 5 a.m. the day of the fire, and went immediately to his office, a small frame enclosure in the receiving room, located in the basement. He lighted the gas in his office and threw away the match. At about 5:18 a.m. another employee discovered the fire in the timekeeper's office. The waste-paper basket, chair and desk were burning briskly.

The fire did only moderate damage in the receiving room but spread to a tile enclosed shaft containing two elevators and eleven small dumbwaiters enclosed in wood. The dumb-waiters and elevators served the dining rooms from one kitchen on the eighth floor.

Equitable Building before the fire was a composite of five buildings, equivalent to ten stories high.

The shaft had openings directly to each story except the fourth. It is claimed that some of the openings had rolling iron shutters. They were evidently open or defective. The other openings had wood doors. The flames quickly extended throughout this shaft and entered the upper stories almost simultaneously. In the shaft it did not communicate with the lower floors; the shaft acted as a chimney, the draft being inward at the lower levels and outward at the top. The fire traveled fastest through the large area of the club rooms and kitchen, which were almost without sub-divisions by even ordinary partitions and contained much combustible trim and furnishings. Probably within forty-five minutes after its origin the fire had spread over most of the main building above the fourth floor and was rapidly working its way down to the lower stories through numerous light wells and other unprotected floor openings. In approximately twenty minutes more the first (east) collapse of the floors occurred, followed thirty-five minutes later

Post-fire view shows severity of burnout and collapse.

by the second or general collapse extending along the Broadway and Cedar Street fronts. In about four hours the fire had destroyed practically everything combustible in the west half of the building above the grade floor and similarly everything in the east half above the fourth floor. It was held in check in the lower stories at the east end and was under control in the remainder of the building about 9:30 a.m. By noon it was mostly out, although burning quietly in some inaccessible portions of the building until late in the night.

The weather conditions were very unfavorable. When the fire started the wind was blowing from the northwest at an average rate of 37 miles an hour. The velocity increased and during most of the fire averaged approximately 55 miles an hour, rising to occasional maximums of 65 to 68 miles. The temperature was 35 degrees above zero Fahrenheit at the start of the fire and fell rapidly, being less than 20 degrees F. before the fire was under control.

Although the wind was from the northwest the tall buildings across Broadway deflected it through Trinity Churchyard, so that it struck the building on the Broadway front from a westerly direction.

There were few persons in the building at the time of the fire so it gained access to the upper floors without being discovered. The fire was thus enabled to spread to an extent where it was beyond control before any concerted effort could be made to restrain it.

The number of unprotected floor openings and large areas, unbroken except by non-fireproof partitions — notably in the restaurants — favored the rapid spread of the fire.

The building was equipped with two 4-inch standpipes, one on each side of the building near the main elevators, supplied by two tanks of 2,250 gallons capacity each, also by siamese fire department connection on street. Each standpipe had a 2½-inch outlet with 100 feet of 2½-inch unlined hose on each floor. The standpipe tank was filled by a 3-inch pipe from two duplex steam pumps. The capacity of one was 350 gpm and the other 450 gpm. Each took suction by a 4-inch pipe from city mains. These pumps had also a 3-inch connection direct to the standpipe system.

After the fire was discovered, about 5:18 a.m., employees stretched hose from the inside standpipe equipment, but on attempting to use it found the hose was too short. Considerable time was lost in providing an adequate length of hose. A policeman, when notified by one of the employees of the cafe, investigated the fire and at once called to another policeman who turned in an alarm from a street box at 5:34 a.m. It is estimated that approximately sixteen minutes elapsed between the discovery of the fire and the receipt of the first alarm.

The first engine company to arrive took a hose line to the basement by the winding stairway leading from the sidewalk. The fire in the timekeeper's office and receiving room was promptly controlled, but it was later discovered that the flames had communicated with the dumb-waiter shaft. It was not then apparent to what extent the fire had spread beyond the shaft and the firemen went up through the building to investigate. As the shaft was somewhat inaccessibly located with respect to the only stairway, it took an appreciable time to determine the full extent of the fire and the necessity for a second alarm.

Eight companies entered the building and operated approximately twelve streams on the second, third, fourth, and fifth floors for periods ranging from fifteen minutes to one hour. Some of these streams were taken off the inside standpipe which was supplied by the private pumps in the basement and a steamer connected to the outside siamese coupling.

All members of the fire department were ordered out of the burning building about 6:35 a.m., shortly before the first collapse occurred, Battalion Chief

Walsh being the only member who did not escape. Later during the fire a number of streams were operated on the lower floors at the east end, mostly by means of ladders to windows.

Scarcely any attempt was made in the early stages of the fire to fight it with hose streams directed from the level of the street. Such an attempt would have been futile, owing to the narrow streets and high wind. The streams were mostly directed from the upper stories of the high buildings opposite across narrow streets, by hose lines taken from the standpipe outlets on various floors. The standpipes were usually supplied by a local fire pump and one or more steamer hose lines secured to the siamese connection at the street. A number of streams operated from roofs of Nassau and Cedar Streets. Two water towers were used. A few streams were directed from the street level in Broadway, but these were evidently used chiefly to wet down the ruins in the vicinity of the vaults.

The effect of the high wind was to blow the hose streams into a fine spray, and at some points streams were not effective at seventy feet — effectiveness of the hose streams from the standpipes was also greatly diminished by the wind. It affected the high pressure streams from the separate fire main system less than those from the steamers.

The nearest hydrant of the high pressure service was approximately eight hundred feet from the part of the block where the fire originated. This placed the service beyond the limit of distance within which it is ordinarily considered as available. It was used later when the operations of the fire department were extended to Nassau and Cedar Streets, the nearest hydrant being about five hundred feet from the intersection of these streets.

The high pressure service supplied eight hose lines, seven of which were 3-inch hose and one was 3½-inch. They were attached to two hydrants. Two of these lines supplied the mast nozzle of Water Tower No. 2; two furnished an equal number of streams from a roof; two were attached to a siamese standpipe connection on a building; one was connected to Water Tower No. 1, but not used; one line was carried to a second floor.

Fifteen lengths of fire department hose burst during the fire. Fourteen were 3-inch hose and one length was 2½-inch. This is not considered significant in view of the amount of hose in use and its frozen condition.

The temperatures developed in this fire seemed to be slightly lower than in other so-called fireproof building fires. Glass and brass fixtures in the upper parts of the rooms, such as electric fixtures and the blades of electric fans, fused freely (estimated about 1,600 degrees Fahrenheit). In the dining rooms and kitchens silverware was fused, as were also brass door knobs and speaking tubes. A brass standpipe valve on one of the upper floors showed incipient fusion. No cast-iron or steel parts were found to be fused (steel melts at about 2,450 degrees F.; cast-iron at 2,200 degrees F.). The fire was probably hottest around the four light shafts on the Cedar Street side of the main building and in the basement at the point where the insulation burned off the telephone and telegraph wires.

Loss of Life.

As far as is known, six persons lost their lives in the fire.

Three employees of the cafe, whose escape was cut off by the fire, jumped from the northwestern corner of the roof to Cedar Street and were killed.

Battalion Fire Chief Wm. Walsh was killed in the building near the main stairway on the 4th floor by the collapse of the floors near the stairway.

Two watchmen lost their lives at the Broadway and Cedar Street corner of the building when the floors at that location collapsed.

It is fortunate that the fire occurred at a time when the building was practically unoccupied or the loss of life might have been far greater, due to inadequate facilities for escape and the early collapse of the cast-iron columns.

For nearly a century, fire departments have been concerned about the transportation and storage of munitions, dynamite and other explosives. Originally, the transportation of such explosives came under the jurisdiction of the Interstate Commerce Commission, but the function of that organization have been assumed by the U. S. Department of Transportation. The explosion described in this article was a well-publicized incident just prior to the U. S. entry into World War I. A delayed alarm of fire, inadequate spacing between munition storage facilities, and the placement of railroad cars loaded with munitions close to buildings were very important elements of this disaster.

The Black Tom Island Disaster[14]

Preeminent in interest among the numerous fires and explosions that have resulted from the present abnormal traffic in munitions of war, stands the Black Tom Island disaster of July 30, 1916. The preeminence consists not alone in the pecuniary value of the property destroyed — though that is estimated at about $20,000,000 — nor in the loss of life and personal injury inflicted on the general public — though nearly sixty people are known to have been direct sufferers (including six killed) — but in its revelation of a condition of affairs under which innumerable human lives and vast quantities of property are menaced every day by the possibility of similar disasters at a variety of points where munitions are being handled in transportation or storage.

The attempt to secure investigation by the Interstate Commerce Commission into the facts and circumstances surrounding the fire and explosion have elicited the fact that cars loaded with millions of pounds of high explosives may be kept in railroad yards in close proximity to buildings occupied as human habitations, at the same time that the storage of much smaller quantities of explosives in properly constructed and approved magazines is governed by strict regulations prescribing the distances which must intervene between the explosives and various classes of buildings, highways and — grim irony — railroads. It has been held in the courts that municipal authorities lack power to regulate the transportation of explosives through

cities, the Federal authorities possessing sole jurisdiction over interstate commerce. Instances are, in fact, on record of the disregard of precautions demanded by a local fire department. On the other hand, the Interstate Commerce Commission has taken the view that in the absence of evidence of violation of its regu'ations governing the packing and handling of explosives, it has no authority to undertake the desired investigation.

The story of the *results* of the explosions — the rain of shells upon the surrounding country, the destruction of huge warehouses and their contents, the damage to the immigration station on Ellis Island, the bespattering of the Statue of Liberty with shrapnel, the strewing of miles of streets with broken glass and shattered signs — has been told in the daily press. The story of the *causes* of the disaster has been a matter of controversy from the beginning and the real facts will probably not come to public knowledge until an official investigation is made by the Interstate Commerce Commission or until and if litigation ensues, but it seems clear that, as a matter of fact, the fire started shortly after Saturday midnight among some freight cars on the pier several hundred feet distance from the warehouses. It would also appear that explosive-laden barges were at or in close proximity to the piers. The first explosion occurred at about 2:08 a.m., but the fire appears not to have gained much headway until after the second explosion which occurred twenty-two minutes later, which seems to have widely scattered burning embers which, apparently, started fires in many places.

There seems to have been considerable delay in informing the Jersey City fire department of the outbreak of the fire. In the official statement issued by the department on the day of the disaster, it was announced that the battalion chief in charge of the fire engines dispatched to the scene on the first call found the fire already a large one, and that he experienced difficulty in getting his engines within striking distance. Moreover, water could not be obtained from the railroad standpipes. From another source it is learned that fire had been observed a considerable time prior to the summoning of the fire department, and the suggestion has been made that the fire had been in progress for more than an hour before the occurrence of the first serious explosion.

Thirteen newly sprinklered brick storage warehouses, owned and operated by the National Storage Company, and several piers leased by that company to the Lehigh Valley Railroad, were destroyed together with more than eighty loaded cars. The warehouses had an excellent private water supply, but the shock of the first explosion is thought to have put the system out of commission. At all events, the only supply available was from a one-pipe line from Jersey City, and this proved utterly inadequate even when aided by fire tugs from the harbor. The discovery of munition boxes drifting about the harbor, into which they had been hurled by the explosions, caused much anxiety among pilots of harbor craft. On the following day a serious outbreak occurred among the ruins caused by the first fire. On this occasion also there were numerous explosions of shells, and the fire chief of Jersey City was reported to have expressed the opinion that the ruins would continue to smoulder for a week. The fire actually burned for fully two weeks before it was quenched.

Notwithstanding the fact that the Storage Company and the Railroad have been censured by a coroner's jury for their neglect to take adequate precautions to safeguard the public in connection with their handling of explosives, the situation does not appear to have been remedied, and large numbers of cars loaded with high explosives continue to be received into the yards of Jersey City. Clearly, thorough and impartial investigation is urgently needed not only to dispel the obscurity which still surrounds the origin of the Black Tom disaster, but to devise effectual means for the protection of the people against any repetition of the offence.

Famous Institution Fires

This huge building was used as a multiple-facility caring for soldiers, handicapped people and children, many of whom were bedridden. Combustible interior finish and the large number of helpless children were important factors in the loss of life.

The Grey Nunnery[15]

A portion of the fifth floor of the west wing of the Grey Nunnery, in Montreal, was destroyed by a fire occurring in the early evening of February 14, 1918, resulting in the loss of life of fifty-three babies.

The home was a large stone building, occupying an entire city block, and part of it was occupied by convalescent soldiers. There were about one thousand persons in the building — two hundred returned soldiers, many nursing sisters, nuns, crippled people, aged and children. Although many of them were stretcher cases, all adults were taken out of the building without loss of life. Several of the returned wounded men, however, were seriously injured by falling debris before they could be reached by the large army of rescuers. A dozen or more had to be carried from the second and third stories on beds.

The fire started on the top floor, near the tower, and immediately ignited the curtains of a near-by window, from which it spread rapidly throughout the wooden interior. The returned soldiers who were not confined to their beds lent their efforts to the rescue work. When the first fireman arrived the soldiers were already at work, at great risk to themselves, in handing children down the fire-escapes. Nearly the whole upper floor was then ablaze. Fire fighters rushed in and seized children right and left. A sudden gust of flame and smoke which burst from the tower made it impossible to reach children lying in their cots in that part of the building, but only upon the definite orders of the fire chief were the soldiers debarred from attempting further rescues.

All of the top floor was occupied by babies in cots, some of them only a few days old. These were the infants lost, all the older children, who were in another part of the building, being saved. When fire fighters, after valiant work, got the fire under control, all the southern part of the top story had been destroyed and a great deal of damage done to the remaining part of the west wing. After the fire was extinguished thirty-eight little bodies were taken from the debris, and on the following day fifteen more were found.

Flimsy, combustible construction, a delayed alarm and difficult removal of patients accounted for severe loss of life in this hospital. Employees tried to stop the fire with hand extinguishers but their efforts were fruitless.

Oklahoma State
Hospital for the Insane [16]

The Central Oklahoma State Hospital for Insane, at Norman, Oklahoma, was partially destroyed and thirty-eight inmates were cremated in a night fire which occurred on April 13, 1918.

The fire was discovered by the night watch in Ward 14 and by an inmate in Ward 15 (second floor) about 3:45 a.m., in or near a linen closet which was near the middle of the building on the first floor. Three wards and the general dining hall were totally destroyed with their contents; thirty-seven of the inmates of Ward 14 and one of Ward 15 were killed in the fire. The value of property burned was estimated at between $40,000 and $50,000.

The cause of the fire is unknown, some thinking it might have been caused from electric wiring, but there was nothing to substantiate the supposition except the absence of any other apparent cause to which it could be attributed. No electric lights were installed in the linen closet in which fire is supposed to have originated. A slight lightning storm was brewing at the time of fire, but no serious electrical disturbances were noted at or previous to the discovery of the fire. Upon discovery of the fire a general alarm was sounded by a steam whistle at the hospital power plant, and all ward watchmen, nurses and other employees proceeded to fight the flames with hand chemical extinguishers and one hose stream. Every effort was made to first get the inmates out of the burning and exposed buildings. It was claimed by the Norman public fire department that they were not called until about 4:15, or until after the fire had been in progress probably thirty minutes. During the fire a moderate wind was blowing from the southeast, and, but for the good work of the public department with the pumping engine, probably several other buildings would have been destroyed.

From information obtainable, in would appear that the fire was fought as satisfactorily as could be expected in view of the fact that removal of the inmates in the wards destroyed was a very slow and tedious procedure, due to their very inferior mental and physical condition. Due to the flimsy and rapidly combustible construction of buildings involved, the flames spread rapidly. It would appear that quite unnecessary delay was experienced in turning in the alarm to the public fire department, but no information was available as to the cause of this delay.

In the wards in which the fire originated and in which all the lives were lost, there was a total of 86 inmates — thirty-seven on the first floor and forty-nine on the second floor — one inside and one outside stair were available for exit from the second floor. The inmates of the lower floor being of the very lowest mentality, quite a large number of them were gotten out only to rush back immediately into the burning building. It was stated that one night watch was stationed in each ward, but no approved clock service was employed and no record was obtainable on effectiveness of this service. Regardless, the fire claimed thirty-eight victims.

This fifty-year-old building had three serious fire experiences before the final disaster occurred. As in other institutions, many of the patients were helpless or violent and most of them reacted in panic to the fire situation. The combustible interior, paraffin wax on mops, and other fire protection weaknesses helped create a tragedy.

Illinois State Hospital [17]

A fire which occurred during the evening of December 26, 1923, in a group of joisted frame buildings, part of the Illinois State Hospital for the Insane at Dunning (just outside of Chicago), caused the death of eighteen persons.

The past history of this institution is interesting. In 1912, the superintendent had called attention to the serious conditions then existing. On August 23, 1911, one of the buildings had been destroyed by fire. On January 17, 1912, another building burned to the ground. This building, built in 1870, had been condemned in 1908, but was forced back into use without remodeling because of overcrowded conditions. On May 4, 1914, another building was burned. On October 16, 1916, two barns on the grounds were destroyed by fire. On December 11, 1918, the tuberculosis ward, with 400 patients inside, took fire. The patients were rescued with great difficulty. After this fire the ward was rebuilt the same as before.

The scene of the fire was a group of six one- and two-story frame buildings communicating through one-story frame passages 30 to 50 feet long. The central building was occupied as a dining room on the first floor and a dormitory above. The area was approximately 5,500 square feet. There was a two-foot unused space under the lower floor.

The passageways were one-story frame with an approximate area of 6,200 square feet. The annexes were used for dormitories, each provided with wooden bins in which the patients' clothes were stored, and small enclosures in which mops, brooms and cleaning materials were kept.

The buildings contained about 300 patients, mostly epileptic, but some violently insane kept on the second floor over the dining room. Many of the patients had received Christmas cigars and cigarettes and they were allowed to smoke in the buildings without restriction. The mop closet in each dormitory contained polishing mops made of six-inch pipe wrapped with discarded bed linen and blankets. Paraffin wax was rubbed on these mops, which were used for polishing the floors. The buildings were heated by steam, the radiators being covered with wire mesh and sheet metal guards.

A private hose reel with 300 feet of hose was kept in a frame shed located within 150 feet of where the fire started. A city hydrant was approximately 300 yards distant. The nearest city engine company was 3½ miles away. No watch service was maintained. A few 2½-gallon soda-acid extinguishers were provided.

The fire was discovered at 5:40 p.m., when all the patients were at supper in the dining room, by a patient who habitually stayed behind to say prayers. It started in the mop closet in Annex No. 5. The cause has not been definitely established, but it is probable that spontaneous ignition of the mops or defective electric wiring was responsible. The patient immediately shouted "fire" and a nurse in the dining room telephoned to the operator in the Administration Building, who sent the alarm to the power house, where a siren was sounded. A city fire alarm box located in front of the Administration Building was pulled at 5:41 p.m. The fire spread rapidly and the patients in the dining room were hustled out of the building and lined up outside. There were only eighteen guards to handle the patients, some of whom were violently insane and all of whom were panic stricken. The doors of the dining hall opened inward, creating a jam, and the lights went out, adding to the confusion. A number of the patients eluded the guards and ran back into the building to save personal belongings. Most of these burned to death. A caretaker and his wife, who had rooms on the second floor over the dining room, went back to get their eight-year-old son. They remained to pack up some belongings and all three perished.

The guards marched the patients to the Administration Building. On the way the violently insane patients bolted and many escaped into the city. Three men broke out of the line and ran back into the building and locked themselves into a lavatory on the second floor. Policemen entered and broke down the door and were able to force the patients down ladders to safety.

The private fire brigade had a fairly good hose stream playing on the fire when the city department arrived, but the buildings were all aflame and some walls were already falling. The wind had been blowing from the southwest and sweeping flames directly toward the other wards, but fortunately at this time it shifted and the firemen were able to save Annex buildings one and two. Seven engine companies and four truck companies responded to the fire. The roads had been transformed to deep mud on account of a heavy rain and great difficulty was experienced in moving the apparatus. Fourteen bodies were found in an area of approximately 300 square feet at the rear of the dining room.

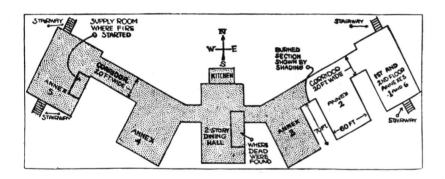

Here is another tragedy in which complete ignorance of fire potential, or deliberate negligence, contributed to the deaths of twenty-five people. Would this be possible today?

Dance Hall Holocaust[18]

On the evening of November 22, 1919, twenty-five persons lost their lives in a Ville Platte, Louisiana fire which destroyed a two-story frame building, the upper story of which was used as a dance hall, while the first floor comprised a grocery and restaurant of doubtful reputation, a clothing store and a motion picture theatre. The fire had its origin in the grocery and restaurant. It appears that the wife of the proprietor was warming some coffee at a kerosene oil stove when the tank ran dry. She instructed a boy employee to fill the tank, which he did from a can at the back of the store. When he reattached the tank to the stove, she again lighted the latter, but the burner immediately exploded, scattering burning oil over the stove and floor. She took a broom to beat the fire out, while the boy fetched a bucket of water, emptying it over the broom, which by this time was on fire. The water spread the flames, and the boy, becoming terrified, ran out of the building. A high school pupil who was also present testified that he pulled the stove away from the partition by which it stood. A hole about five inches in diameter had been burned in the partition wall, and several men coming out of the card room endeavored to extinguish the fire by a further use of water pails. Remembering that his mother was in the picture theatre next door, the boy went to fetch her, and found smoke already coming through the partition wall. The people in the theatre were advised to leave quietly, and all of them escaped.

In the meantime the fire, working its way up inside the partition wall, burned through the floor of the dance hall about in the middle of the building.

At once there was a rush for the only stairway, and the panic-stricken dancers found themselves jammed in the stairway and confronted by an equally panic-stricken crowd which was madly fighting its way up in an insane attempt at rescue. The result was a scene of indescribable horror; many of those who escaped were only able to get out by walking on top of those who were already jammed in the stairway. Finally the staircase collapsed under its human burden, causing the partition wall which separated it from the clothing store to give way. Those who were not already dead or badly injured could then escape to the street.

The building burned to the ground, the efforts of the local fire department having no effect upon the fire. Although there seems to have been considerable delay in sending in the alarm, the department was severely criticized for its slowness in getting to work. It was alleged that, even after the delay, if the department had shown a reasonable degree of efficiency, the blaze might have been controlled without loss of life.

The facts that a dance hall should have been permitted in the upper part of a highly combustible structure, the lower story of which was occupied for mercantile purposes and the display of motion pictures, that the hall should have been left without any other means of exit than the inadequate stairway on which the "jam" occurred, and that, with such conditions obtaining, so little thought should have been given to the provision of adequate public fire protection, form a sad commentary on the indifference of the American people to the value of human life.

Searching for the dead.

Ruins of the dance hall.

147

Another chapter was added on Christmas Eve to the long story of schoolhouse tragedies. Thirty-two men and women and children lost their lives when flames swept this rural school. Nearly every element favorable to fire was present. It is hardly conceivable that conditions could have been worse, even in so rudimentary a structure as this one-story building. Here is how the United Press reported this tragedy on Christmas Day, 1924.

The Hobart School Fire[19]

Christmas of 1924 dawned as a day of wailing and weeping for Babb Switch, seven miles south of Hobart, Oklahoma.

Horror-stricken families huddled in groups around the blackened embers of a country schoolhouse, in which thirty-two men, women and children of Babb Switch gave up their lives Wednesday night, December 24.

Rescue workers beat through the charred timbers Thursday, seeking other bodies.

Thirty-seven wounded were in Hobart hospitals Thursday. Many of these, it is feared, will die.

More than half of those who died in the flames were children. Identification of the cremated bodies in many cases was impossible.

A mass meeting of the countryside was called in Hobart to decide upon funeral and relief arrangements. Officials in Oklahoma City and Tulsa were preparing to send aid.

The Miles sanitarium and the Physicians and Surgeons hospital in Hobart were filled to capacity.

Dozens of the living bore burns and bruises as mementoes of the nightmare night.

The youth and aged of Babb Switch were gathered in the little frame schoolhouse Christmas Eve.

"Fire!"

There was a hush, an ominous rumble. High up on the spangled Christmas tree, close to the glittering Star of Bethlehem, a candle had toppled into tissue-papered gifts.

Those nearest rushed to put out the flame, sweeping the tree.

Too intent on their task they struck the tree to the floor.

There was panic and stampede. Mothers fought to reach their children. Men shouted for order.

The flames leaped to the ceiling, and in the glare and smoke the happy garb of Saint Nicholas was snatched aside, revealing the gaping figure of Death.

There was only one exit from the frame building. It was jammed with shrieking, fainting men and women.

Victims trampled upon each other. Women guarded their children with their own bodies. Men strove heroically — and vainly.

Survivors could not remember today all of the things that happened in the doomed schoolhouse during these hectic minutes.

The teacher of the little school was among the missing and survivors told a story of seeing her battling amid the flames to get her children to safety.

One of the men nearest the door stood at the exit pulling his neighbors to safety until finally bodies became so jammed across the sill that he was helpless.

Those who gained safety tore frantically at the wire covering of the windows of the building. Sleet and snow had stiffened the wire like tempered steel.

Some sped for barn-yard axes. But before they had returned the scythes of Death already had made axes futile.

More than 200 persons were in the schoolhouse when the flames broke. The room measured only 25 by 36 feet.

Hardly a family in the little community was left untouched by the catastrophe.

The injured lay in the snow banks surrounding the school, while the uninjured vainly fought the flames.

Later automobiles were turned into improvised ambulances, and the injured were carried across sleet-covered roads to Hobart.

A telephone call to Hobart was the first intimation of Oklahoma's worst Christmas catastrophe. Ambulances and fire apparatus were hurried to Babb Switch.

The fire broke out in the schoolhouse just as Santa Claus, ruddy and cheerful, had begun handing out presents to the children gathered there.

It was the second fire disaster for Oklahoma in the last few years. At 3 a.m., April 13, 1918, fire broke out in the hospital for the insane at Norman, and thirty-nine patients lost their lives.

Nearly every precaution for safety to life and against fire was neglected in the building where this appalling tragedy occurred. The newspaper account and the accompanying pictures make most of them apparent. Some of these were:

1. The building itself was of the lightest kind of frame construction — the least fire-resistive type of building.

2. For illumination gasoline lamps were provided. During the first moments of the fire these

exploded, accelerating the already swift progress of the flames throughout the wooden crematory.

3. Open flame candles swung from the boughs of the tinder-dry cedar Christmas tree. Here was a hazard so well known that it is astounding that no one present recognized the danger. This was the fatal hazard.

4. The building was crowded beyond reason. Reports set the number of people in it at between 200 and 250 persons. Even 200 in a room 25 by 36 feet is several times what any law would permit.

5. The one door opened inward. It was a veritable "check-valve," making it next to impossible for a surging crowd to pass through. The state fire marshal had vigorously condemned such construction in schools, but could not act because not backed up by any building code.

6. But one exit was provided to the building. The windows, which might have been used for emergency means of egress, were nearly as heavily barred as a prison. These barriers were placed at the direction of the school board to keep campers from breaking into the building and to protect the panes from youthful ravagers. They were deliberately and securely bolted in place.

7. Not a single provision to take care of a fire should one start had been made. There was not even a fire pail. The blaze was attacked in its incipiency by wraps and bare hands, and was probably spread rather than checked.

8. Finally, those who had escaped from the building had to stand by and see the fire burn unhindered, for there was not a hose line or even a drop of water in the vicinity.

So far the list is of factors that had a bearing on this particular fire. Two others perhaps bear mention, as either might have been a contributory factor:

9. A frame shed was built adjacent to the school building as a garage to house the teacher's car.

10. The building was heated with an ordinary stove. The overheating of a similar stove had destroyed the schoolhouse once before. On that occasion teacher and pupils escaped.

This is a partial list of the points that were neglected. There were probably others which have not come to light. Any one of the above in itself was serious. With all of these hazards, it is no wonder that the flames were able to do their worst.

The Hobart School was similar in construction and size to this rural school, also in Oklahoma.

A close view of one of the steel window screens which resisted all efforts to tear them from the windows of the burning school, and which were held in part responsible for the large loss of life.

Once again, the fallacy of a "fireproof" structure is dispelled by practical fire experience, as described in this 1926 report. (*Refer to page 136—Ed.*)

The Second Equitable Building Fire[20]

The fire in the Equitable Building, New York, occurring at 3:05 a.m. on February 16, 1926, for which four alarms were sounded, occasioned great interest on the part of the public as well as of the fire protection fraternity. The building was of modern fire-resistive construction and few would have imagined that a fire of any magnitude was possible in a structure of this character.

This fire recalled vividly the spectacular fire of January 9, 1912, which destroyed the old office building of the Equitable Life Assurance Society at the same location. The fire of 1912 destroyed the building, which at that time was supposedly a "fireproof" structure and much concern was felt as to the safety of the present structure. It speaks well for the modern structure to know that a severe fire was confined to one floor of a cut-off fire section, and so far as actual fire damage was concerned, to a single suite of offices on the thirty-fifth floor. There was, however, some smoke and water damage to other portions of the same floor and to the floors above and below, and a heavy fire and water damage to the pipe shaft where the fire originated and water damage to the lower floors of the building. The fire in the file room on the thirty-fifth floor adjacent to the pipe shaft was very severe and destroyed practically all of the contents of that room excepting contents of two safes. The damage to the structure itself is comparatively slight. The work of the fire department in controlling and extinguishing the fire, described by Chief Kenlon as "the highest the world has known," is worthy of the highest commendation. The rescue of the workmen trapped in the penthouse forty stories above the street was thrilling and it is remarkable that no fatalities or serious injuries were recorded.

The building is a superior type of fire-resistive construction, including incombustible trim and floors throughout, with standard protection at the main floor openings, and is divided into four nearly equal fire sections above the first story by eight-inch brick walls with standard automatic fire doors on one side of the wall at all openings. Automatic sprinklers are installed at a few places to protect special hazards, but no sprinklers were installed at points involved in the fire.

The standpipe system consisted of four six-inch standpipes in stair shafts with 2½-inch outlets on each floor and four on roof with reducing couplings and 150 feet of 1½-inch standard hose at each outlet. The standpipes are supplied by eight-inch pipe from two 12,500 gallon tanks on the 41st floor, with a total of 10,000 gallons reserved for fire purposes. Standpipes are cross-connected at 36th intermediate and second sub-basement floors; reducing valves are located on the 7th, 18th and 28th floors.

The Pipe Shaft.

Extending from the engine room in the second sub-basement up through the roof structure to the roof is a 12- by 12-foot pipe shaft. Three sides of the shaft are constructed of 6 inch hollow tile blocks, while the fourth side, which is an enclosing wall of an elevator shaft is 8 inch brick. The shaft is entirely open to the second sub-basement. At the top it is roofed over except for a hooded metal vent. At intervals of three to five floors there are door openings leading to the floors, protected by hollow metal fire doors with three point latches.

Prior to the fire, one of the 6 inch cold water supply lines loosened from its hangers on the grid iron beams and slid several feet downward in the shaft. Two crews of plumbers were put to work to make repairs in the lower and higher portions of the shaft. These men entered the shaft from the third floor doorway and from the 35th floor doorway. The preparatory step taken by the workmen was to strip the pipe of its insulating covering. This consisted of paper and felt and an outer canvas covering (about one inch over all) and was piled upon the grid irons at the floor levels as the work progressed. The crews employed the use of portable cord connections carrying electric bulbs and it was the practice to use the stationary light on the particular floor where they were working for general illumination and connect the portable cord connections to an outlet on either the floor above or below, or both, for concentrated light for the work in hand. A short circuit developed in one of the cords in use by the lower crew when they were at the 8th floor level, about midnight, but it is claimed it was detected and a new cord substituted.

150

Gasoline torches were to be used in the repairs but it is claimed that the work had not progressed to the point where they were necessary and that the appliances had not been taken into the shaft. Examination of the torches found in the basement indicates the statement to be true.

The Story of the Fire.

According to the testimony brought out by the investigation, the fire was discovered by the men in the lower portion of the shaft who were then working at the third floor level. It was described by one of them as originating from a flash and when first detected appeared to be at about the first floor level. One of the men descended the shaft to notify the engineer in the second sub-basement and in so doing passed the fire at the point of origin. This man was evidently very much excited at the time, for he cannot state what it was that appeared to be burning or give any definite details as to exactly what happened. One of the remaining men of the lower level crew drew two pails of water, which he threw at the fire, without result, and the crew left the shaft at the third floor level and closed the door to a locked position. The upper crew were trapped in the shaft and climbed up the ladders to the upper story of the penthouse, where they left the shaft by the doorway, which they closed, smashed a window in the penthouse and climbed out on the narrow ledge, from which they were subsequently rescued by the firemen. It is stated that they clung to the ledge for at least a half hour. As previously stated, the upper crew entered the shaft at the 35th story and did not close the door to a locked position. This type of door locks itself upon closing by a mechanical device in the form of a striking plate on the edge of the door, which throws the bolts into place in the door frame. Examination of this door shortly after the fire showed that the bolts had not been thrown, that the door had not been pried open and that it had been left open when the men entered. The doorway opened directly into the file room of the offices into which the fire entered.

The fire originating below the level at which the men were working might lead to the theory that perhaps they had been smoking and dropped a lighted match or cigarette upon the flammable material beneath them. They steadfastly claim that they were not smoking and strengthen their claim by saying that because of the heat in the shaft, due to the various steam pipes, which made it uncomfortably warm, they had no desire to smoke.

The fire having started almost at the bottom of the shaft found plenty of material in the form of the paper and felt pipe covering piled upon the grid irons and the other pipe coverings to feed upon, and aided by a strong natural draft, it quickly spread up the shaft. Temperatures were reached by the burning material to cause twisting and expansion of the five-inch uncovered gas line. The expansion subsequently ruptured the line at the 10th floor level and dropped the lower section of the pipe at the break about 19 inches. That this pipe had parted was not known until the day after the fire. Failure of gas flow from the piping in the Club on the upper floors caused an investigation by the management and it is stated they removed some 65 or 70 gallons of water from the pipe and the meter. The water discharged into the shaft by the fire department during the progress of the fire, collected in sufficient quantity in the open end of the lower broken gas pipe to cause a water seal, extinguishing the blaze and preventing further flow of the gas.

After the engineer was notified by the workman who went down the shaft, an attempt was made to extinguish the fire by the local fire brigade. It is stated they first visited the 3rd floor door opening in the shaft, but were unable to open the door because it was expanded by the heat. They then went to the 15th floor and being unable to open the door, broke a hole in it and inserted the nozzle of a 1½-inch hose from the building equipment. They then went to the 24th floor and were also unable to open this door. The fire in the shaft at this time had gained great headway and they decided to call the fire department. From the time the man went down the shaft to notify the engineer until the alarm of fire was turned in, it is estimated that 12 minutes had elapsed.

Upon the arrival of the fire department, practically the entire shaft and the quarter portion of the 35th floor were enveloped in flames. The standpipes of the building were utilized to extinguish the fire. The regular water supply of this equipment was augmented by connecting lines from the high pressure system. The fire was under control in about one hour after arrival of the fire department.

Thirty-seven youngsters died in this orphan asylum which did not have sprinklers, was surrounded by heavy snow which blocked the approach of fire apparatus, and featured open stairways and refuse chutes that permitted rapid spread of fire and intense smoke. Confusion in exit procedures caused many children to lose their lives, and a young girl made the maximum sacrifice.

"The Victims Were Children" [21]

The Hospice St. Charles in the City of Quebec was visited by fire on the evening of December 14, 1927, with the loss of thirty-seven lives. The victims were children, the majority of them being little girls between five and nine years old.

The Hospice St. Charles was an orphan asylum consisting of a main building built in 1831 and an annex consisting of two adjoining parts built in 1920 and 1925. At the time of the fire there were three hundred and seventy-one children asleep in the various dormitories and about twenty sisters in charge. While this institution was designated as an orphanage, it served also as a school, and the three hundred and seventy-one children included many with living parents whose frantic efforts to find their children added to the confusion and intensified the horror of the catastrophe.

The exterior walls of the building were of stone. The interior was of wooden joisted construction. The central portion was four and one-half stories high and the two wings were three and one-half stories high. With open stairways and open refuse chutes leading from the various floors to the basement, the interior construction of this building favored the rapid spread of fire. There were three wide stairways, and two iron fire escapes were provided on the rear of the building.

Between 8 and 9 p.m. one of the sisters made an inspection of the building and found everything in order, with no indication of smoke or fire. The exact time and point of origin of the fire are not definitely known, but it is clearly established that the fire started in the basement and there is a strong probability that it originated in a room which was used for storage purposes.

According to custom all children and the sisters had retired at 9 p.m. About 10:45 one of the sisters

was awakened by the crackling sound of burning wood and she immediately rushed to give the alarm, which consisted of ringing the assembly bell. Others were apparently awakened by the smoke about the same time, and the testimony indicates confused shouting and cries of fire throughout the building.

The sisters marshaled their charges in the several dormitories and started them out of the building. Had it not been for the good discipline which apparently prevailed it is probable that the loss of life would have been much larger. The smoke filled the building rapidly and the corridors and stairways were soon impassable. A large number of those rescued were taken down ladders by the fire department; some jumped or were thrown from the windows and were caught in life nets. It is not clear how many escaped by the different exits, but there is definite testimony that one of the fire escapes was used effectively during the early stages of the fire.

Apparently the greater part of the loss of life occurred in the section of the building directly over the point of origin of the fire. One of the sisters testified that many of the children from five to nine years old who lost their lives had returned to certain death in the dormitories after having been started toward the exits. There is a report that as one group of children marched from their dormitory they were met by a cloud of smoke from which they recoiled and retreated to their beds, shortly to be suffocated there.

The heroine of this fire, Rose Anna Gaudreault, was a young girl of seventeen who lost her life in an attempt to rescue the youngest of the children of whom she had charge. When last seen she was mounting the stairs to go to the children on the third floor. Nearly two hours later her body was found in a corridor on the third floor, near the entrance to a room where several little girls were found dead on their cots.

The first alarm for this fire was transmitted from a box located in front of the main building. This box was "pulled" by a passer-by at 10:48 p.m. A second alarm was turned in by a policeman at another box at 10:55 p.m., and a general alarm was turned in by a deputy chief of the fire department at 10:56 p.m. On arrival of the first section of the brigade, heavy smoke was found issuing from the basement windows, also flames from the roof on the southwest side of the building. Lines of hose were immediately laid from hydrants about 250 feet distant and carried through the yard at the rear of the building to the basement. Water was stated to have been turned on without delay, but the use of the streams was abandoned at the outset, firemen turning their entire attention to the work of rescuing the many children reported to be confined in the building and who at first were understood to have been safely removed.

Men responding to the second alarm also united with those already engaged in rescue work. Nine or ten ladders were unshipped from trucks by this detachment and thrown over the high iron fencing surrounding the building grounds. This unusual procedure was due to the fact that trees and heavy accumulations of snow prevented trucks from being readily driven into the grounds, and was the cause of considerable delay in erecting ladders against the building. Two aerial trucks, for use on the central or higher portion of the building, subsequently entered the grounds through the main gateway but were only suitably placed after some trees had been cut down.

About the time the general alarm was transmitted two lines of hose were laid directly from a hydrant located in the grounds opposite the main entrance of the building. These lines were carried up ladders and water directed through windows on the upper floor. Attention was next directed to laying additional lines directly from near-by hydrants and the placing of pumping engines. With engine and direct hydrant streams in scrivce, the fire was fought from the street level, roof of the kitchen at rear, taking lines of hose into the second story of the building and up ladders raised for the purpose. During the course of the fire the temperature was moderate and there was little or no wind.

Indications are that most of the victims who lost their lives were killed by suffocation rather than by burns, although after the fire it was found that many of the bodies were so badly burned as to make identification difficult. Most of the children died in the building, but two of those rescued alive died later in a hospital. The coroner's jury in its investigation of the disaster returned a verdict of accidental death, but added to its verdict a demand "that more effective supervision be required during the night, and that automatic sprinklers be installed in all buildings of this kind."

The characteristic process of a flammable dust explosion is a sequence in which the dust is first thrown into suspension, then ignited by some source. The first action of throwing the dust into suspension may come from an explosion, or simply the jarring effect of dropped machinery or some other object, so that dust moves through the air. This mine tragedy explains the classic sequence of such an explosion.

The Kinloch Mine Disaster[22]

Forty-six men were killed and four were injured in an explosion and fire at the Kinloch Mine, Parnassus, Pennsylvania, on March 21, 1929. The explosion originated underground, and it was here that forty-five lives were lost. It extended to a structure above the mouth of the mine, which was seriously damaged by fire, and one death and the four injuries occurred here. Two hundred and fifty-eight men were underground at the time of the explosion; fortunately, due to partial rock dusting, the explosion was limited and two hundred and twelve escaped on the day of the explosion and one the following day.

The general arrangement of the mine slope is shown by the accompanying diagram. The workings were approached through a slope pitching at an

angle of 30 degrees. This opening was of sufficient width to accommodate a conveyor about five feet wide, a car track for hauling mine refuse and supplies, and a footway.

The mine was ventilated by an exhaust fan, which at the time of the explosion was exhausting 241,500 cu. ft. of air per minute. Of this, 160,400 cu. ft. per minute, pure air from the surface, entered by way of the slope, making a velocity of 1050 ft. per minute over the conveyor.

The mine was electrified, with mine machinery electrically operated, and electric lights. Most of this was not of a "permissible" character for a gassy mine. ("Permissible" means approved by the U. S. Bureau of Mines, following tests.)

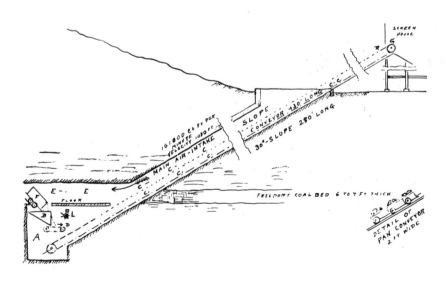

The Explosion.

At 7:25 a.m. on March 21, the loaded strand of the conveyor broke a few feet below the top. A portion of the conveyor lodged on the incline above the slope portal and the remainder went to the bottom with its load of 50 to 60 tons of coal. The sudden impact of the conveyor at the bottom threw a dense dust cloud into suspension. This was ignited by a spark from electric wiring or equipment deranged by the impact. The dust explosion extended to the surface and through a large area of the mine with considerable violence.

All of the men in the mine who lost their lives were involved by the force or effect of the explosion from the moment of its occurrence except seven, who attempted, after having been otherwise directed, to make their escape by passageways leading to the slope. Workmen who were fortunate enough to be able to travel and act promptly after the explosion soon found their means of escape to the slope opening cut off by afterdamp, and were compelled to find their way to openings to the surface through unmarked and unfrequented passageways, portions of which were partly filled with water. These openings were from two to four miles from the slope portal.

Temporary repairs having been made to the ventilating fan structure, and men and procedure organized for entry to the mine by way of the slope portal, and the last group of workmen having arrived at the surface, the ventilating fan was put in operation. In a few minutes after the fan was put in operation, trained men and mine officials descended the slope and began exploration and recovery work, which was continued until the mine was fully explored and all bodies recovered. Recovery operations were considerably hampered by the finding of five fires. Three were extinguished by direct methods, but the other two could not be reached because of smoke and other hampering conditions and were therefore sealed off.

The explosion traveled up the slope to the Tipple Building, which was ignited and burned for three or four hours. One man was burned to death here and four were injured. This structure, as shown by the accompanying illustration, was entirely of steel construction, with no combustible material except coal dust and pieces of coal, some wooden flooring, and the paint on the corrugated iron walls. The damage to this structure was largely due to the fire rather than to the explosion, which presumably did not have great force when it propagated outside the mine and into the Tipple Building, the floor of which was 25 to 30 feet above the ground.

Mine Explosions

During the past 100 years thousands of people have been killed and many more burned or in-

jured in major and minor explosions that have occurred in mines. Most of these accidents are attributed to ignition of gas and dust, particularly in abandoned rooms and old workings. The U. S. Department of Interior Mining Enforcement and Safety Administration in Pittsburgh, Pennsylvania is the principle authority for maintaining safety conditions and standards of operation in mining properties.

A review of mining statistics indicates the seriousness of hazards in mines. Between 1891 and 1900 there were thirty-eight major explosions resulting in 1,006 fatalities. In addition, the U. S. Bureau of Mines listed 250 minor explosions for the same period, each resulting in one to four deaths, with a total of 426. Three-fourths of these minor explosions and fatalities occurred in anthracite mines. Yearly fatalities between 1911 to 1920 averaged 279, almost a third less than the four hundred-plus average for the previous decade. In the next ten years (1921–30), fatalities averaged 320, but in the following two ten-year periods between 1931 and 1950 the averages dropped to 102 and 86. In 161 major explosions between 1881 and 1910 6,133 fatalities were recorded. In the next thirty years, between 1911 and 1940, 7,000 fatalities were recorded in 223 major explosion disasters.

One of the recent mine disasters was at the Sunshine Mine in Kellogg, Idaho, May 2, 1972. This was a silver mine which also produced antimony and copper. A fire broke out about 3,400 feet underground and ninety-one miners died. Other major mining disasters included: Monongah, West Virginia, December 6, 1907 — 362 killed; Dobson, New Mexico, October 22, 1913 — 263 killed; Scofield, Utah, May 1, 1900 — 200 killed; Coal Creek, Tennessee, May 19, 1902 — 184 killed; Eccles, West Virginia, April 28, 1914 — 181 killed; Cheswick, Pennsylvania, January 25, 1904 — 179 killed.

The London, Texas, School Disaster[23]

A gas explosion that caused the death of 294 school children and teachers resulted in one of the major disasters of the present century. It destroyed a high school in the Texas oil field district about 3:15 p.m. March 18, 1937. The site of the explosion was the unincorporated community of London, about four miles east of the little town of Overton in the northwest corner of Rusk County, ten miles from Henderson, Texas. The community consisted principally of the school property of the London Independent School District, Inc., comprising the large high school building, several smaller buildings and seven active oil wells owned by the school district. Beyond the school property on all sides were oil well derricks, gas flares, and a distant refinery, all typical of the oil field industry. Because of the wells on its property it had the unique distinction of being about the wealthiest rural school district in the world, catering largely to families of the oil field workers over a territory as much as 15 miles distant in several directions.

The blast occurred with the suddenness characteristic of such explosions, although with some unusual features. Every witness agreed that there was but one explosion and that it was a low rumbling noise, with none of the blast or roar that might be expected. Yet there is evidence of a most terrific force in the great extent of devastation and loss of life that came almost instantly; testimony of bodies

tossed 75 feet into the air; an automobile 200 feet distant crushed like an eggshell under a two-ton slab of concrete that had been hurled from the building. And as a further evidence of the terrific force, the established point of origin indicates that the explosion had to break through an 8-inch concrete floor slab before starting on its path of destruction. Many of the children in rooms directly above this point were literally blown apart or mangled beyond recognition.

There was very little of the structure left standing after the blast and most of that had to be pulled down to prevent falling after the debris was removed, which was practically overnight. From the earliest photographs taken, the blast seems to have completely demolished the entire front section, 58 by 254 feet, both the north and south wings back to the east 25-foot additions, and the center auditorium wing back to the last 40 feet of the east end, which had to be razed when the supporting wreckage was removed. All that remains of a $300,000 modern school plant is a badly wrecked two-story addition forming the rear of the north wing which will have to come down and a few broken walls where a similar addition completed the south wing.

The high school was erected in 1932 with additions in 1934; it was built on ground sloping in two directions, with a down grade from north to south

When a great many people are killed in the sudden violence of an explosion, the survivors, and the general public, at first find the incident beyond comprehension. Many witnesses, friends, and relatives of the dead and injured undergo various stages of shock and disbelief and it may be some time before they accept the reality of the incident. Much of the nation reacted in this manner when the lives of teachers and children were extinguished in this famous Texas school explosion.

and from west to east. The main section was 58 feet wide and 254 feet long, fronting west, with classrooms on each side of a 10-foot central corridor. At the north and south ends, wings extended east to a depth of 136 feet, terminating in two-story "T" additions. Directly in the center the two-story auditorium section formed another wing extending east 115 feet. The whole plan formed a letter "E" and covered nearly 30,000 square feet of ground area.

There were two rooms devoted to laboratory and science work, one having 12 laboratory tables, each of which had two ½-inch gas pipes carried up from under the floor with two bunsen burners on each line. These two rooms were located in the front of the main section near the north end of the building. A manual training shop was located in the lower floor of the north wing; it contained several machines driven by individual motors and one portable sander with a heavy cord and plug switch for a wall socket.

The original plan called for a central steam heating plant, but was abandoned before construction started because of the higher cost as compared with individual gas stoves. A common type of gas-steam radiator was adopted after the School Board members had consulted several people presumed to be familiar with the subject, including members of other school boards. The device had the appearance of an ordinary steam radiator, but was an in-

dividual heating unit comprising a gas burner at the base, under a small water chamber cast into the unit. Steam circulated through the hollow sections of the radiator and heats by radiation like the standard steam type. The burner was partly enclosed, but communicated with the outer air through a circular hole about an inch in diameter where a match could be inserted to light it. The unit was equipped with a small regulator at the gas supply end by which pressure could be adjusted, and had a safety valve or blow-off on the water chamber. It was a well-known make used extensively in the Southwest and was considered as safe as any gas heater on the market. It was used to a large extent without vents, but could be equipped with a vent pipe for a wall flue if desired. There were seventy-two of these heaters in the building and as there appeared to have been some belief that the state school authorities required all classroom heaters to be ventilated, some effort was made to vent only those heaters located in classrooms. A short piece of threaded pipe extended from the vent outlet on the heater into the wall, neatly finished with a collar at the wall entrance, giving the appearance of a properly vented unit, but as there were no flues in the walls this supposed vent simply entered a hole punched into the wall tile.

Room ventilation was mainly through windows on all sides, although two or three small wall vents were located in the rear two-story section. An important weakness in ventilation, however, was the totally inadequate provision for venting the concealed dead space under the main section of building where the destructive gas is presumed to have accumulated. This continuous concealed space had a superficial area of over 15,000 square feet entirely enclosed by a 12-inch concrete wall, but having communicating doors opening into classrooms of both the north and south wings. The depth of this space varied from two or three feet to as much as six feet at some points. A conservative estimate of three-foot average depth means about 46,000 cubic feet of dead space for which only four small 12- by 24-inch vents were provided. Two of these vents were at the north and south ends, 254 feet apart, and two on the east wall where the distance and angle to the opposite ends could not give proper circulation.

Gas Fuel and Pipe Lines.

A great deal of publicity was given to the fact that the school officials had discontinued the use of commercial gas from the utility company that had been costing as much as $250 a month, and had tapped a free gas line of "wet" or "residue" gas from a gasoline extraction plant without permission. This is a point that requires some explanation of a common practice that the school officials adopted with good intention.

When an oil well is brought in, it flows under its own gas pressure. This gas is separated from the oil and piped off to some distant point and burned in a flare. It usually has a heavy gasoline content and in some locations gasoline companies set up extraction plants to which this gas is piped, the gasoline extracted and the residue gas piped back to the lease from which it came. This procedure follows a standard form of contract; the gasoline company has no authority to sell or dispose of the residue gas, neither can they give permission for anybody to tap the line, although the gas is headed back to an open flare to be wasted. There appears to be an obligation on the gasoline company to serve notice of disconnection should they "find" the line tapped, but as a matter of inclination and general custom they are willing to look the other way until the line is covered. It is therefore a common practice for churches, schools and even private homes to get this free gas if within reach of the pipe line. While the gasoline has been extracted, it is still known as "wet" gas; it is more or less unstable as to volume, and pressure fluctuates considerably; it contains impurities that are removed in commercial gas. Apart from the inconvenience of keeping the heating units adjusted to the irregular supply and pressure, there is no particular hazard in its use as compared with ordinary commercial gas which has caused similar explosions.

This gas supply reached the building through a 2-inch pipe into a gas regulator of standard commercial type located outside the south wall. It had a capacity of 50 psi on the high pressure side reduced to 5 ounces on the domestic supply side. The average pressure on the extraction plant residue line was said to be 25 psi. From the regulator the 2-inch line entered and extended in a full circulating loop under the entire concealed space of the main section of the building, suspended by metal straps from the concrete slab forming the first floor. Numerous 1½-inch lines extended up to the room heaters from this line. Other 2-inch lines were carried to the rear of the two-story wings and extended through the attics, with drop lines to the heating units on the two floors below. As all gas pipe involved in the area of explosion was blown away, it was impossible to trace the leak or leaks that undoubtedly existed. It developed, however, that all repairs and extensions to gas or water lines, electric wiring, etc., were made by one of the school janitors who were "jacks of all trades" and probably masters of none. They may have neglected to test the pipes for pressure leaks before turning the gas on; failed to pull the threaded joints up tight; omitted to apply lead or other compound at the joint, or even cracked a pipe without knowing it.

There were rumors that the explosion was due to nitroglycerine or dynamite. Dynamite was being used at the athletic field in the construction of a running track. One of the teachers testified that when he heard the explosion he thought it was another shot at the field. It was also found that eighteen sticks of dynamite were stored in a lumber room under the auditorium stage and went through the explosion intact. By the process of elimination it became obvious that the cause of the explosion was the ignition of a large pocket of gas in the large improperly vented space under the floor. This theory was corroborated by evidence of the heavy concrete floor slab above the space being blown up from below, and testimony that bodies, desks and lockers were found on the bare ground, indicating that the slab floor had been blown out before the bodies and other objects fell.

It appeared most probable that the gas was ignited in one of the rooms where laboratory or shop work was conducted in the northwest corner of the building. The shop, where most of the electric devices were used, adjoined and communicated with the under floor space at this point. Progressing further on this theory it appeared that there should be two classes of dead; those in the immediate path of the blast, who should show evidence of burns and mutilation, and those killed by falling wreckage. This line of investigation developed the fact that all bodies with evidence of burns or with arms and legs blown off and missing came from classes in the northwest rooms and that the worst case of burns was the teacher who was just starting a class period in the woodworking shop. One of these shop pupils who escaped with slight injury testified that the door into the adjoining concealed space was about half open, that the electric switches and wall sockets were within two feet of the open door on the same wall, and that the teacher was in the act of plugging in a portable connection to the sander when the flash and explosion followed.

This practically established the fact that the large under floor space of the main section was pretty well saturated with an explosive mixture of gas and air from a source of gas leakage that will never be determined. Some of this gas mixture passed the open door into the shop and was ignited by an arc formed when the two prongs of the portable switch touched the socket before it was driven into place. The resulting flash followed the gas back under the building, creating the superheated gas and pressure necessary to blow the building up. While the northwest corner of the building was blown out first, there was no appreciable interval in the collapse of other portions. The blast traveled the entire 254 feet of building instantly and literally blew it to pieces. Witnesses testified that the roof was lifted at one corner at the same time that walls were blown out at another corner.

The theory of released gas from the oil fields floating into the premises was discounted by the fact that the building is located on high ground, while floating gas would be confined to the lower levels. The U. S. Bureau of Mines engineers drilled all over the school campus and found no seepage of gas from any underground source. The question of using the residue or wet gas had no bearing, since any leaking gas under the same circumstances would produce the same result. The much publicized theory of accumulated gas in the hollow tile walls was started through ignorance of the conditions. There was no way in which free gas could enter the walls. Furthermore, it was not a wall explosion. The question of faulty gas heater vents was discussed at the court hearing as well as in the press comments and while the whole question of the heater installation is subject to criticism, it had no bearing on the explosion.

The only point of attack on the conclusion of accumulated gas in the under floor space was the testimony of one of the janitors that he had entered that space at 10 o'clock that morning, five hours before the explosion, and had lighted matches to smoke and find his way about in the dark; and indicating that there was no gas leaking at that time. This is conceded to be possible, as his point of entry was at the extreme south end of the building and his travel distance not over 50 feet, according to his own evidence. He was, therefore, about 200 feet away from the area where the explosion originated and gas might have been leaking at the north end of the building without reaching an explosive mixture 200 feet away. Furthermore, with the great amount of gas pipe and fittings up to two inches in size, it would have been possible for a leak to

start after his departure, and form an explosive mixture in much less than five hours.

No fire followed the explosion, presumably due to the small amount of combustible material. The main structure was of concrete steel and tile as previously described. The windows were metal factory sash. Apart from the interior wood trim at the doors and the furniture, everything was practically noncombustible up to the wooden roof deck. This roof deck was covered on the outside with tile and was cut off from the interior by metal lath and plaster ceilings.

While early reports put the total casualties at a larger figure, the final count showed that the deaths totaled 294, including 120 boys, 156 girls, 4 men teachers, 12 women teachers, 1 woman visitor and a 4-year-old boy visitor.

There is no record of the total number of persons in the building at the time of the explosion, but it appears that the majority of the occupants were killed or injured, and that practically all of those in the main part of the building were killed instantly or were fatally injured.

Those who survived were in the wings, mostly on the upper floor. There were many reports of jumping from windows, which appears probable, as the windows were not more than 10 feet above the grade. There were three boys who were in the shop where the explosion originated who escaped with slight injuries. They were in the far end of the room and were blown into the rear addition.

There were many reports of persons who survived by reason of protection of desks and other objects of furniture. It is probable that some children and teachers crawled under desks, but it appears doubtful that there was enough time for any teacher to direct an entire class to get under their desks. The survivors who were immediately able to get out were assisted from under the wreckage in conditions varying from slight shock to broken limbs or fatal injuries. Many were pinned under the wreckage for long periods and were released only when heavy material was lifted by the use of jacks.

Loss of the Hindenburg Zeppelin[24]

The largest loss fire of 1937 occurred on May 6, when the German Zeppelin *Hindenburg* burst into flames near the tail while landing at Lakehurst, New Jersey, at the end of one of her transatlantic flights. Once started, the flames spread over the 803-foot ship at a rate estimated at 50 yards per second. Fabric that was thin enough to constitute almost no obstruction in the fall of an occasional wrench or other tool dropped by a mechanic working inside the great silver hull of the ship, as shown by small holes where such tools had torn through, did not offer any slow-burning qualities to prevent the ship from being totally consumed in just a few short moments. Some of the many competent observers at the scene noted that the destruction was completed in exactly 32 seconds. In 32 seconds, the once moot question of the safety of passenger transport by lighter-than-air craft was swung far toward the negative.

This was the 119th and the largest airship built in Germany. It, and all but one of its predecessors, was filled with flammable hydrogen. Ten others, planned but not built, were also designed for use with this gas. It is difficult to understand how the safe passage of 7,063,000 cubic feet of hydrogen in a bag subject to all the hazards of nature could be expected for even one trip. Of the 118 previous Zeppelins six mysteriously disappeared, 21 were dismantled, 18 were surrendered or destroyed during the war,[*] and 71 were involved in accidents of one sort or another. The now long grounded, helium-filled Los Angeles, and the long lived Graf Zeppelin are the only two now in existence. This is an impressive record of failure, and yet the extreme foresight, constant vigilance, and understanding of the terrific hazards by her German masters brought the *Hindenburg* safely through ten round trips across the Atlantic. That same thoroughness has made it possible for the old *Graf*[**] to cheat nature in eight years of safe flying. Sparks from her engines have trailed harmlessly into space behind her for all that time without once encountering proper conditions for igniting an escaping wisp of hydrogen. The *Los Angeles*[***] worked out her usefulness under the comparative security of inert, non-flammable helium. Her sound construction was proof against the structural failures that dumped other helium-filled ships into the rubbish pile.

The *Hindenburg* was not under such a lucky star. A spark from her engines, a static spark, perhaps a hot exhaust pipe found the proper concentration of hydrogen, and 36 lives and $3,750,000 in property disappeared.

[*]World War I
[**]*Graf* Zeppelin, a predecessor of the *Hindenburg*
[***]A U.S. dirigible

In recent years, scientists and transportation authorities have recommended use of dirigibles as a means of transporting freight and other supplies at reasonable cost. During the 1920s and 1930s such airships were used commonly by several countries, particularly Germany and the United States. The sudden, startling tragedy of the German Zeppelin *Hindenburg* ended the use of Zeppelins for the next forty years. The fact that hydrogen was a cheap, lighter-than-air gas offered some economy; but its flammability and the light fabric enclosing the Zeppelin created a different problem.

Natchez Dance Hall Tragedy[25]

One of the worst fire disasters of the twentieth century occurred in a small dance hall in Natchez, Mississippi, a quiet old river town of about 16,000 population. There, on the evening of April 23, 1940, 207 persons lost their lives and 200 others were injured when fire involved combustible Spanish moss over the dance floor. There was only one main exit and the small windows apparently were nailed shut. The panic stricken people piled up at the rear wall of the building, most of them dying from suffocation or from being trampled.

The Rhythm Club in Natchez was used as a place of entertainment by the black people. It was only one story high with the roof and sides made of corrugated iron. It was a long, narrow structure, measuring 120 feet in length and thirty-eight wide. The main entrance opened to an interior lobby which in turn opened onto the dance floor. At the rear of the building was a bar, with the orchestra platform in the opposite corner. Along the side and rear walls were approximately eighteen windows, all boarded to keep out "gate crashers." There were no skylights or upper windows to vent the heat of any fire which the metal walls of the building would confine like an oven.

On the night of the fire, a popular Chicago orchestra had attracted a record crowd and about 700 or more persons were inside the building at about 11:15 p.m. The fire started near the hamburger stand at the front of the dance hall when the gray Spanish moss, hanging from ceiling joists, ignited and began to burn rapidly. People in the lobby in front of the dance floor were able to get out the door quickly, but others were trapped by a hedge of fire when the burning moss dropped and ignited clothing of the people. In a surging movement the hundreds of victims within the hall pushed back to the rear, where most of them died. It was surprising that so many escaped since the only main exit passed through a cloak room and the lobby where the doors opened the wrong way, inward.

The fire department responded to an alarm within minutes and the fire was quickly extinguished. Sheet metal siding was torn off to allow rescuers to reach the victims but many, trapped below the deck, suffocated before they could be released.

Like every other fire tragedy, this incident presented lessons which should have spurred many cities into improving their fire prevention regulations. The Rhythm Club had one main exit which was partially blocked. Doors leading to this exit opened toward the interior; but people trying desperately to escape expected the doors to open the other way. The mass of highly combustible Spanish moss was an obvious danger; so were the closed windows. The outer iron covering of the building, while noncombustible, confined the heat with maximum intensity on the people trapped inside.

Similar fires, in earlier years, should have served as a warning. In 1929 in a club in Detroit, the combination of combustible oak leaves hanging from the ceiling and boarded windows caused twenty-two fire deaths and twice as many injuries from panic when a carelessly discarded match ignited the decorations.

In 1936 nine persons lost their lives in a New York City restaurant when fire spreading through a window from the floor below flashed over festooned silk cloth ceiling decorations. Today, fire codes require that no furnishings or decorations of an explosive or highly flammable character shall be used in any place of assembly or other occupancy. But, after Natchez, other fire disasters would underscore the need for such common-sense fire safety.

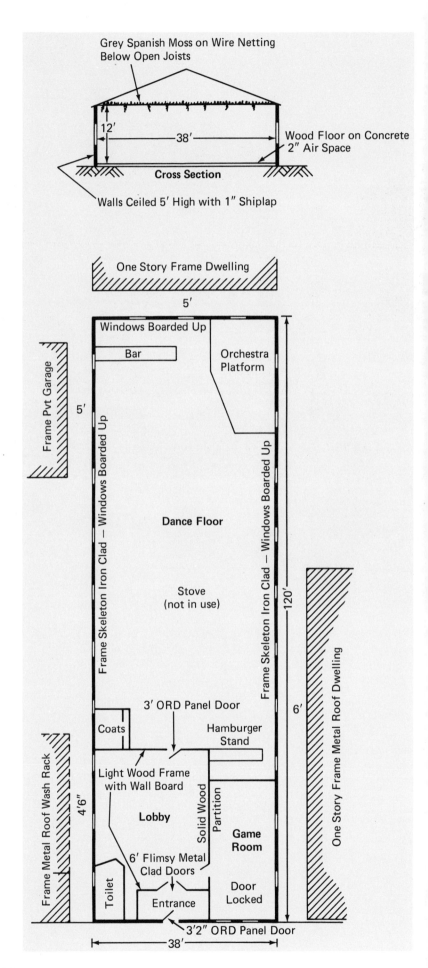

163

Scene at the Naval Air Station, Pearl Harbor, Hawaii on the morning of December 7, 1941 when a surprise attack by Japanese aircraft devastated U.S. ships in the harbor. Thousands of servicemen and civilians were killed and injured. (Official U.S. Navy photograph.)

1941-1945

Chapter VI

During World War II fire was used as a weapon of destruction more than in any previous conflict between nations. For the first time in history, incendiary weapons could be dropped by the thousands from aircraft, or directed with reasonably good accuracy by drone aircraft, or rockets that traveled fifty, a hundred, or more miles to their targets. A million or more human beings died as a consequence of deliberately planned fire, and countless others suffered injury. When that war ended with the dropping of atomic bombs on the Japanese cities of Hiroshima and Nagasaki, people throughout the world were thoroughly shocked at the destruction caused by those bombs. They had not yet been informed of the far greater destruction and human suffering caused by the deliberate, strategic utilization of incendiary fires created by smaller weapons.

It is worthwhile to review some realities of life in the United States just before the war began, because everyday customs, available materials, manufacturing processes, and capabilities for fire protection and fire control were highly important in some of the catastrophes that occurred.

Most aircraft belonged to the military, the Navy, and other government agencies, although privately owned planes and commercial aircraft were increasing. All were propellor-driven, single- or double-winged planes. The turbo-prop and the jet engine had not yet been developed. Larger cargo and passenger planes had up to four engines with variable pitch propellors. Gasoline was the principal aircraft fuel. Only the major cities had airports large enough to serve as terminals for transcontinental or overseas flights; most airports were small, with unpaved runways, and little fire protection. Airport crash trucks and fire brigades were minimal, if available at all.

The marine industry was fairly well-developed. Passenger and merchant ships were powered by coal or oil. Most ships were of steel construction, although some wooden-hulled ships were still in service. Past lessons of fire-at-sea had helped to stimulate safe practices of fire protection aboard ship and dockside, but the shipping industry had much to learn.

In 1940 most U. S. single family dwellings had indoor plumbing facilities and community water supply, but many did not have these conveniences. Wood, coal, oil, and gas were used for heating and cooking, but the all-electric home was only experimental. Dwellings were very susceptible to kitchen stove fires, particularly from "range oil" burners, usually supplied from a large glass jug of kerosene fastened on a metal support immediately adjacent to the stove. The fuel flowed by gravity from this jug, through a length of flexible metal tubing, then into the stove interior to the burner. If too much fuel was forced into the burner while the wick was ignited a flash fire or explosion could result. This occurred frequently and loss of life from these fires was frequent.[2]

Few buildings had central air conditioning or other means for moving air for cooling. At the time it was not appreciated that such equipment would increase the spread of fire products — heat and toxic gases — but in warm weather noncooled buildings could be intolerable for sustained working effort; it was logical to improve these conditions.

Railroad trains, the principal means of transportation for passengers and cargo, were used extensively. From Boston to New York was a trip of five hours, best done overnight. From New York to Chicago was thirty-six hours, or more than two days, depending on a number of possibilities. Cross-country was at least a five-day trip coast-to-coast, also dependent upon different circumstances. With such slow transportation and few cross-country highways the nation had to move the greatest portion of troops, war materials and civilians in the atmosphere of full mobilization for war.

Few plastics or other synthetic materials were available for mass distribution and use. Clothes and textile furnishings were made of protein (animal), cellulosic (vegetable), or mineral materials but man-made fabrics were becoming available. Thermoplastics such as nylon and rayon were beginning to be made in quantity when the war started. Natural rubber was common for shoe heels, auto tires, rainwear, cushions, as a sealant of openings, as material for toys, and for hundreds of other uses.[3]

Radio was the principal means of instantaneous communication to the general public. Television had been demonstrated at the 1939 World's Fair in New York City but would not be available for nationwide usage until the late 1940's. Some fire departments had two-way radio installed in cars or fireboats, but these departments were few. On the fireground, communication to the alarm center was by telegraph or telephone.

Public attendance at theatres, dance halls and night clubs was consistently high; the defense buildup in the late 1930's helped relieve the massive economic and social depression of the years since 1929, and there was great demand for entertainment.

Hospitals, small in those days, had fewer medical specialists and patients. Oxygen was being used, but not to the extent it is used today. Some anesthetics were highly flammable or explosive. Otherwise, hospitals and institutions had serious fire protection deficiencies, flammable interior finish, open stairways and undivided areas, and inadequate alarm, extinguishment, and exit facilities.

Cities and other large communities had public water supply, usually fed from a reservoir, lake or stream, but many small communities had to rely on well water, consequently fire fighting in these communities had to be accomplished with little or no hydrant supply.

Fire departments mostly used straight stream nozzles for interior and exterior fires, and wore filter-canister masks or self-contained breathing apparatus when smoke and toxic gases were unbearable. Injuries and fatalities were great among men who did not have this equipment for protection.

At that time, the tallest building in the world was New York's Empire State Building which was destined for a very unusual fire.

Then, early in the morning of December 7, 1941, Japanese aircraft bombed Pearl Harbor in Hawaii and the United States began active participation in World War II.

Honolulu Air Raid Fires[4]

Eyewitness report by the late Chief William W. Blaisdell who commanded the Honolulu Fire Department at that time.

Surprise was the word for it. While certain plans to be followed in the event of an attack on this city had been made, more were in a formulative condition.

Within half an hour after the first attack, all off-duty firemen were called in by radio broadcasts from three stations, and all but a few who live in rural Oahu had reported within two hours. Scores of volunteers thronged the stations, and workmen from other city departments were assigned as emergency firemen.

Thus it was possible to use each of our units as a company. In addition to this, nine commercial trucks were commandeered; each was loaded with 1,200 feet of 2½-inch hose and placed in strategic location, each in charge of a trained fireman, whose position was filled by a volunteer. This was also done with our one reserve pumper and a Reo truck with booster equipment.

During the 72-hour period following the first attack, one unit only was dispatched to each alarm. Boxes which came in were not transmitted, all assignments being made by telephone. During the height of the attack, no effort was made to mop up, it being felt that, immediately after a fire had been extinguished enough to prevent its spread, apparatus should be returned to quarters in readiness for other fires.

As near as we of the Fire Department can tell, there were no incendiary (Thermite) bombs dropped, all of them being demolition.

Necessarily, our efforts were directed toward the best protection possible for the high value district and the waterfront. Three of our companies had been sent to Hickam Field where fires had been started in hangars, barracks and planes on the landing field. Hickam's apparatus had been put out of service by bombs.

The men of these three companies stayed with their lines amid a hail of machine gun bullets from raiding planes and shrapnel from large demolition bombs. The hydrants were rendered useless when the main was broken by a bomb and our three pumpers drew water from the bomb crater, pumping through 6,400 feet of 3-inch and 2½-inch hose, the last apparatus returning to the city at midnight.

Each piece of apparatus bears holes from bomb fragments and bullets. One pumper was put out of commission when machine gun fire demolished the ignition switch, and had to be towed back. One hosewagon was set on fire, presumably by an incendiary bullet which pierced the gasoline tank. The motor, however, was not badly damaged and this piece will be put back into service soon, as repairs are being rushed on the body. The engine was put back into commission the same afternoon.

Our greatest loss and the one most keenly felt was that of two captains, one with 34 years' service and the other with 23½ years, and a hoseman who had been in the department a little over three years. These men were killed by machine gun bullets from dive bombers and may God hasten the day when those treacherous Japs shall pay for this. Six other firemen were injured, one a lieutenant; one has returned to duty

and the five are still in the hospital recuperating from wounds in various parts of their bodies.

On December 7th we answered thirty-nine alarms. Not all were caused by bombs, and not all the bombs that hit buildings started fires. The largest, in the semi-residential district, involved thirteen buildings before it was brought under control with one hose-wagon manned by a captain and six hosemen. This loss will approximate $165,000. The engine which ordinarily responds with this hosewagon confined another fire to the building of origin, scorching the walls and roofs of houses on either side.

Simultaneously with the above mentioned fires, which were only a block apart and both started by bombs from the same plane, a large two story wooden frame school, built in the shape of a U, caught fire, probably from sparks from one or both of the other two fires. Meanwhile a high wind had arisen, which prevented our streams from doing very effective work. Nevertheless this building was saved by the crews from the engine above mentioned, the lieutenant in charge having left two men at the dwelling fire using hydrant pressure and moving in on this fire, in con-juction with another engine which had been dispatched. Using four 2½-inch lines they confined this fire to the second floor and roof of the bottom of the U.

We have recently purchased seven small trucks without body, which our shop is rapidly converting into hosewagons. In the event of another air raid, we have made arrangements for nine other trucks and drivers to report at designated points, together with sufficient volunteers to man them. Our fire wardens (civilian volunteers) are constantly instructing our people how to handle incendiary bombs and continually warning them of the not remote possibility of other attacks.

As for advice to Fire Chiefs in other cities, I can only say this. Prepare for the worst and insist on getting enough equipment on hand to handle the biggest fire you can imagine, in the most hazardous district of your cities.

Debris, dust and smoke covered the water of the bay in Pearl Harbor as the air attack intensified. In background are some of the fires on land which had to be attacked by the fire department, badly handicapped by loss of men and equipment. (Navy Department photo.)

The magazine of the destroyer *USS Shaw* **(DD-373)** explodes in a violent eruption during the Pearl Harbor attack. **(Official U.S. Navy photograph.)**

Fire in the Pacific Theatre[5]

At 6:30 a.m. on December 7, 1941, the auxiliary ship *USS Antares* returning from target practice operations sighted a Japanese midget submarine just outside the entrance of the bay at Pearl Harbor in Hawaii. A radio message was sent immediately to the nearby destroyer *USS Ward* which promptly closed in and sank the submarine at 6:45. A report of the incident was sent to the officer in charge of the naval base at 7:12 a.m.

Meanwhile, on nearby Oahu Island, at 7:02 a.m., an army private at a mobile detection unit noted indications of a large flight of aircraft about 130 miles away heading towards Oahu. At 7:20 he reported this information to his superior officer, a lieutenant at the Army General Information Center. The lieutenant apparently considered the information routine because a large flight of American aircraft were scheduled to arrive at the islands very shortly. Half an hour later the first of 150 to 200 Japanese dive bombers and torpedo planes attacked the U. S. fleet anchored in Pearl Harbor.

Within three minutes this carefully planned attack sank or damaged nineteen American ships, destroyed $25 million dollars worth of aircraft and an equivalent value of buildings, supplies and ammunition, killed 2,383 servicemen and civilians, wounded 1,842, of whom many later died, and left 960 missing in action. Of the eighty-eight ships in the harbor, every battleship and all but a few of the other naval craft were hit by bombs or torpedoes. Fires aboard ship and on land were terrible. Navy physicians in the first sixteen hours after the attack treated 960 casualties, most of them suffering from horrible burns.[6]

After the first attack subsided the Japanese planes circled and returned at 8:40 to strafe and bomb the airfields, barracks and other targets. By 9:45 the attack was finished and the U. S. naval fleet in the Pacific theatre of operations had suffered a grievous setback.

On the same day Japanese aircraft attacked U. S. bases at Guam, Midway and Wake Islands, (the latter including Peale and Wilkes Islands), and bombed Clark Field, a U. S. military airport in the Phillipines, 65 miles from Manila.

The damage to major ships in Pearl Harbor ranked among the highest in the history of naval losses, more than the U. S. Navy suffered in all of World War I. Battleships *Arizona* and *California* were sunk; the *Nevada* was burning extensively and had to be beached; the *Oklahoma* capsized; and the *Pennsylvania* and *West Virginia* were heavily damaged.[7]

Cruisers *Helena*, *Honolulu*, and *Raleigh* were damaged and the target practice ship *Utah* was sunk. So was the minelayer *Oglala*.

Seaplane tender *Curtis* and the repair ship *Vestal* were badly damaged. Destroyers *Downes* and

Cassin were a total loss, and the *Shaw* lost its bow.

Hundreds of Navy crew personnel had to leap into the flaming, oil-covered harbor waters and attempt to swim to safety. The powerful effect of fire as a weapon of war was confirmed once again.

About half of nearly 500 army and naval aircraft lined up closely and neatly at Hickam and Wheeler Fields, at Kaneohe and the nearby Marine base at Ewa were burned or completely destroyed when Japanese dive bombers blasted them with machine gun bullets and incendiary shells.

A few months later the Japanese air forces again relied upon fire when they bombed Manila, shortly after it had been declared an "open city" and was thereby acknowledged as being defenseless. In the last week of December, 1941, this air attack set fire to the large gasoline storage depot at Nichols Field (airport) in Manila, then was directed upon the city for two hours and forty-five minutes. Some 6,000 dwellings were burned or blasted by bombs with countless casualties resulting.

In left foreground is the *USS Oklahoma* (BB-37) with its crew in desperate action to assist the burning 31,500-ton *USS Maryland* (BB-46) in center of picture. At right small boats are patrolling to pick up survivors of capsized ship. (Official U.S. Navy photograph.)

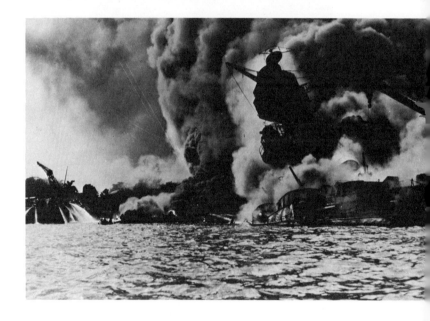

Tragic scene of the *USS Arizona* (BB-39) which like the *USS Nevada* lost most of its personnel who were trapped below decks after bombs struck. (Official U.S. Navy photograph.)

Small boat in foreground is rescuing seaman off its bow while his ship, the *USS West Virginia* (BB-48) burns heavily in the background. Further back is the *USS Tennessee* (BB-43). (Official U.S. Navy photograph.)

Aircraft carriers were highly susceptible to attack by aircraft and huge fires, when their large quantities of flammable liquids became ignited. This is the "Death of the Soryu", a painting by an artist with the surname of Benny. The *Soryu* was one of the Japanese carriers destroyed during the Battle of Midway. (Official U.S. Navy photograph.)

Despite these great losses, within less than one year the U. S. naval fleet was able to engage Japanese naval forces at sea and in the air, and the tide of war changed. Again, fire and explosions aboard ship and on land became highly significant.

On April 20, 1942, a squadron of sixteen B-25 medium bombers took off from the deck of the U. S. aircraft carrier *Hornet* to make a surprise bombing and incendiary attack on the Japanese cities of Tokyo, Yokohama, Nagoya, Kobe and Osaka. Fifteen of the sixteen planes were destroyed in crashes or rough landings; only one reached its designated land base. Many of the crews bailed out over Japan, China or Russia. But the fires they started were burning armament plants, dockyards, railroad yards and refineries and a new phase of the war had begun.

Next came the battle in the Coral Sea, one of the most decisive of the war. It was also the first major engagement in naval history in which surface ships never exchanged a shot; aircraft did all the shooting and bombing. The battle began the morning of May 8, 1942, when a U. S. scout plane spotted a Japanese task force of two carriers, at least four cruisers, and a group of destroyers 175 miles northeast of the American force that included the carriers *Lexington* and *Yorktown*, heavy cruisers *Minneapolis*, *New Orleans* and *Chicago*, and eleven destroyers, plus two Australian cruisers. In the hours that followed the Japanese lost a carrier,

four heavy and three light cruisers, two destroyers, four transports and four gunboats, with a cruiser and destroyer probably sunk and twenty other ships damaged. The American group lost three ships — the carrier *Lexington*, the destroyer *Sims*, the tank *Neosho* and sixty-six aircraft. Personnel casualties totaled 543 in the U. S. ships.

The *Lexington* was ill-fated. In the early moments of the encounter it dodged nine of eleven torpedoes launched at her in as many seconds. But it was hit by the other two and a 1,000 pound bomb wrecked her port forward 5-inch battery, killing the battery crew. Even with this damage, the carrier remained in action to the end, but several hours after the battle, with all her planes on board, a heavy internal explosion occurred while the ship was cruising at reduced speed of twenty knots. Gasoline escaping from ruptured below-deck fuel lines had ignited. The ship stopped as flames enveloped her entire length. The order was given to abandon ship and most of her crew were able to escape by ropes to other ships, or by life rafts, or rubber boats. As the captain who commanded the carrier was sliding down a rope to safety, the torpedo warhead locker exploded and with a final huge detonation, the *Lexington* sank. Ninety-two percent of her crew had been saved.[8]

The battle of Midway occurred June 4, 5, and 6 near the American Midway Island about one third the way between Pearl Harbor and Tokyo. As in the Coral Sea battle, the vulnerability to fire damage of aircraft carriers and other large ships became an important factor.

The first to be hit was the carrier *Kaga*, blanketed with bombs from U. S. planes until she finally exploded. Next the *Akagi* was blasted with bombs and a torpedo rammed into her side. The *Soryu* was completely on fire with gas, blazing above and below decks. In fact, gasoline from wrecked planes on all three carriers burned intensely. Two battleships were hit and burning and one destroyer was sunk before American planes headed back toward their own carriers. The carrier *Hiryu* apparently escaped damage.

In the meantime, the American carrier *Yorktown* was hit by three bombs and large fires started. They were brought under control, but the engine had been crippled and the ship lay dead in the water. However, emergency repairs were made and finally the carrier was taken under tow. Later, however, she was again attacked by bombers burned, exploded, and sank. Before this attack, another Japanese carrier was attacked by American bombers, left blazing from stem to stern, and she sank the following morning. A cruiser and a destroyer were blasted and set afire. Despite the loss of the *Yorktown* the Battle of Midway was considered a U. S. victory — the first decisive defeat suffered by the Japanese navy in 450 years.

On October 26, 1942, the carrier *Hornet* was destroyed in a battle near Santa Cruz Island. A Japanese suicide bomber aimed his burning plane down the carrier's stack, exploding his bombs over the signal bridge. Simultaneously, torpedo planes attacked and the carrier, after dodging fourteen of these missiles, was struck by two, which destroyed all her power and communications. A second suicide plane smashed into the ship and fires raged in many quarters. Even so, they were extinguished in half an hour and the *Hornet* was taken in tow by another ship, only to be attacked again by bombers, after which she sank.

Three other U. S. aircraft carriers in the Pacific suffered great damage by fire in sea battles but survived. On February 21 off the coast of Iwo Jima, the *Saratoga*, oldest and largest carrier in the U. S. Navy, was struck by seven bombs from Japanese aircraft. The entire hanger deck burst into flames and burning planes and leaking fuel oil extended fire throughout the ship. One hundred twenty-three of her crew were killed or missing and 192 were wounded. Despite this devastating experience the fire fighting crew was able to control fires and later the ship was sailed to port.

A month later during the assault on Kure, Kobe and the Inland Sea the U. S. carrier *Franklin* suffered an incredible experience. While lying off-

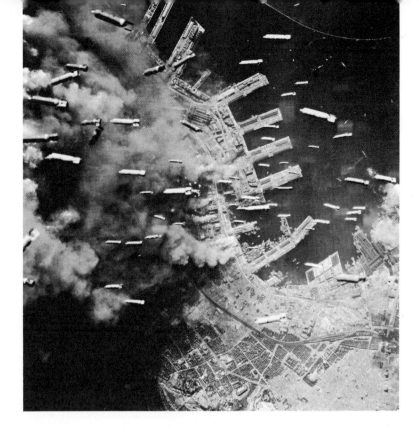

Incendiary bombs dropping on Kobe, Japan in March, 1945. (Official photo U.S. A. A. F.)

shore and refueling she was struck by a dive bomber which dropped bombs from stem to stern. Planes on the carrier deck burst into flames, igniting gasoline drums and ammunition stores and creating an inferno the length of the ship. Again, good fire control action and support from nearby ships helped to keep the carrier afloat, but she was an inferno. About 832 of her crew were killed and 270 were wounded. Exploding ammunition and flammable liquids created a tremendous fire situation, but the 27,000-ton carrier not only stayed afloat but later sailed 12,000 miles back to the Brooklyn, New York Navy Yard, scarred and twisted, but afloat, and she later returned to sea duty.

During the battle of Okinawa on June 21, 1945, the U. S. aircraft carrier *Bunker Hill* was struck by two suicide planes and in thirty seconds was completely involved by a fire that was increased by exploding ammunition. Three hundred seventy-three of the crew were killed, nineteen more were missing and 264 were wounded but the fire was eventually under control and the ship was taken to port. An unusual aspect of this incident was the action of the ship's captain who performed a dangerous maneuver in order to minimize the fire. The decks of the carrier were practically covered by a sea of flaming oil and, while the ship was underway, the captain gave orders to make a sudden turn. This caused the carrier to list sharply and a great deal of the flaming oil fell into the sea, causing diminishment of the topside burning.[9]

Loss of a
Major Resource[10] . . .

It happened in Fall River, Massachusetts, October 11, 1941. The Japanese attack on Pearl Harbor was two months in the future, but in Europe, Germany had overrun all defenses and had knocked England to its knees with a savage air assault. The United States was mobilizing its armed forces and a major part of its industrial effort was the supply of England and other allied countries with ships, aircraft and vitally needed defense materials.

In the Fall River plant of Firestone Tire and Latex Company about 14,000 tons of crude rubber had been accumulated for processing into wartime materials. Rubber was becoming a scarce commodity and this storage represented nearly ten per cent of the entire stock of crude rubber stored in the country as a defense measure. At that time not enough synthetic material was available as a substitute for rubber.

Most of the buildings in the plant area were former cotton mill structures, quite old with plank on timber floors. Most were sprinklered although the sprinkler system was being supplemented for additional protection.

This famous fire started in a dryer on the third floor of one of the buildings. A delayed alarm to the fire department, the closing of a sprinkler valve at a critical moment, an untrained plant fire brigade, and the involvement of the huge quantity of rubber were some of the factors that led to this severe industrial loss. The fire spread to involve twenty-five buildings and destroy their contents. About fifty paid and call fire departments from Massachusetts responded on mutual aid calls between midnight and 6 A.M. the next morning. The spread of fire was stopped by October 12, but the huge stock of rubber continued to burn for another week. At that time it was the largest single industrial fire loss in recorded history, and one of the most serious fires that ever struck the U.S. economy during a state of military emergency.

Dresden, Germany was practically obliterated in this incendiary and explosive attack of March, 1945. It was reported that 300,000 people died in the resultant fire storm. (Official photo U.S. A. A. F.)

Fire From the Sky[11]

Since October, 1939, Britain and Germany had greatly improved the technique of dropping bombs and incendiaries from aircraft. The major cities of both countries had suffered tremendous damage in these air raids and it was obvious that, for both countries, victory meant survival. Both sides used different sizes of incendiary and explosives but Germany, as the war progressed, changed its bomb weapons. For example, when aircraft dropped incendiaries on a British target perhaps one out of ten bombs would be explosive, but in appearance it looked like a typical incendiary bomb. Actually, it was sensitive to detonation if any person tried to move it. Quite often these bombs would land on roofs or gutters without igniting or exploding.

Phosphorus bombs were used to some extent, but they were not too satisfactory. If such a bomb burst in a building, the phosphorus would splash on furniture and other combustibles; as it dried, these would start to burn and extinguishment was difficult. Then there were "butterfly" bombs, which did not explode when they landed but could be set off by the slightest vibration. Bomb squads had to destroy them in place.

The unmanned German "fly" bombs or "V1" missiles also did a great deal of damage and caused many casualties. The sound of their propulsion could be heard from some distance but the worst moment came when that sound stopped and the bomb headed toward the ground. One witness reported seeing five of these bombs come into London within ten minutes.

The V2 rocket was equally bad but its sound could not be heard. These rockets traveled at a rate of a mile per second and carried a 2,000 pound charge in the nose. In those air attacks the British National Fire Service lost about 700 firemen and twenty-five firewomen, killed in action, with about 6,000 injured. In one London raid ninety-one fire fighters were killed and several times that number were injured. A fireman named Frank Eyre wrote this poem to commemorate the experience:[12]

And many learned the nastier ways of dying,
Or limped back, maimed and shattered, from the
  strife,
While all endured unpleasantness and danger
Continually — and learned to like the life.

The first American troops to arrive in England had to adjust to the devastating aerial attacks. It was about April, 1942, that the United States Air Force was able to join the British in retaliatory attacks on Germany.

In the British Royal Air Force and the U.S. Air Force there were two schools of thought concerning the strategic efficiency of air attack. One theory maintained that "precision" attack on factories producing airplanes, oil and other essentials without which the enemy could not wage war, was the quickest and most effective means of attack. The other theory maintained that it was best to kill civilians and wear down enemy resistance through "area" attacks on city targets which were easier, operationally, for the air forces to find and hit. Both techniques were applied extensively, but after a while the RAF preferred to carry out area attacks on cities while the U.S. Air Force applied precision bombing. (Later, in Japan, the U.S. shifted to "urban" attacks because precision targets were difficult to find and hit.)

As the tide of air attack changed, German cities were subjected to day and night bombing from aircraft, and bad fires resulted from almost all these attacks. One report mentioned that thirty-nine German cities had extensive burned out areas; another stated that forty-nine of the principal cities lost thirty-nine percent of family residential units. Berlin, Hamburg, Dresden, Stuttgart, Kassel and Darmstadt developed fires with fire-storm characteristics. In Munich and many other cities, thousands of individual fires were ignited but did not develop to fire storm proportions.

Another lesson to be learned in the battered towns of England and ruined cities of Germany is that the best way to win a war is to prevent it from occurring. That must be the ultimate end to which our best efforts are devoted.

Horatio L. Bond, author of Fire and the Air War

Fire Storms[13]

The tremendous hurricane of fire (from incendiary attack on German cities) caused the air to be drawn toward the fire from all directions with such terrific velocity that it tore trees apart, and prevented firemen from coming close enough to be within range of hose streams. Soon it caused building walls to collapse into streets, and further prevented fire departments from bringing apparatus into the area. In some cases, it prevented them from withdrawing equipment which, as a result, was destroyed, and, in some instances, firemen were killed.

Because of the terrific heat and showers of embers, existing open spaces, even parks, could not be used by the department and they eventually discontinued their efforts to extinguish or hold the fire. They used every means to rescue the thousands who were trapped in basement shelters in the area. The fire department estimated that 55,000 persons lost their lives. Thousands of victims, later found in basement shelters, showed no indication of having been burned, but apparently died from lack of oxygen, heat inhalation, or asphyxiation from coal and other gases. Other thousands, in attempting to escape from shelters down flame-swept streets, were burned to death. Those missing were probably buried in the ruins, or their bodies had been completely consumed by fire. Of the number cited, 45,000 bodies were recovered. The department reported that through their efforts, 18,000 persons were rescued from the fire area.

Damage to property was enormous; 35,700 dwelling apartments were totally destroyed; 4,660 severely damaged. From a total of 450,800 family apartments, 253,400 were destroyed or made unfit for use. A total of 5.9 square miles of buildings was totally destroyed. A larger area was damaged. Many industrial establishments, stores, ships and automobiles were destroyed or severely damaged.

175

Ruins of Hamburg, Germany after incendiary attack of 1943. (Official photo U.S. A. A. F.)

Fire Attack on Hamburg[14]

Professor Graeff, consulting pathologist to the *Wehrkreis X* (military defense area), in Hamburg, gave a very vivid description of the air raids on the night of July 27–28, 1943. The crowded conditions in a city of the size of Hamburg, with its few parks and large squares, the height of the apartment houses, and the age of the dwellings were all contributory factors to the magnitude of the catastrophe.

Soon after the sirens had sounded — a little before midnight on a clear night — the first bombs dropped. The warning was adequate for everyone to go to his shelter or bunker, and thereby evacuate the streets. High explosives and "air mines" destroyed houses, creating craters in streets and courtyards, ruining lighting and the power supply not only in the city at large, but also in the individual blocks, and opening the gas and water mains (no gas escaped from the gas mains). In several bomb craters water accumulated from burst water mains ran into shelters and basements and thereby caused a great nuisance. At the same time incendiary bombs started fires which spread particularly in thickly inhabited parts of town in a very short period of time. Thus in several minutes whole blocks were on fire and streets made impassable by flames.

The heat increased rapidly and produced a wind which soon was of the power and strength of a typhoon. This typhoon first moved into the direction of the fires, later spreading in all directions. In the public squares and parks it broke trees, and burning branches shot through the air. Trees of all sizes were uprooted. The firestorm broke down doors of houses and later the flames crept into the doorways and corridors. The firestorm looked like a blizzard of red snowflakes. More scientifically, firestorm is a mass of fresh air which breaks into burning areas to replace the superheated rising air.

The heat turned whole city blocks into a flaming hell. Those who were still in the streets or for some reason had to leave their homes crowded into a high bunker (a concrete tower shelter) or into a subterranean air raid shelter. Thus the number of people in shelters was doubled and tripled over the number considered safe.

The first serious danger in houses which had not been hit and had withstood explosions nearby became apparent when the lights went out, the water stopped running, and cracks formed in the walls. Air raid wardens on the roofs were threatened by the firestorm and crumbling roofs. In many cases, windows and

176

exits from shelters were blocked by rubble and thus the shelters were safe against fire. As the temperatures increased in the streets from the spread of large-scale fires many of the occupants of the air raid shelters realized the precariousness of the situation, yet very few tried to escape into areas not endangered by fire. In the course of hours the air in the shelters became increasingly worse. Matches or candles did not burn. People lay on the floors because the air was better there and they could breathe easier. Some vomited and became incontinent. Some became tired and quiet and went to sleep. In some shelters oxygen cylinders were available and produced better breathing conditions for at least a short period of time. Wherever the ventilators were still working they brought in hot smoky air instead of cool fresh air, so that they had to be turned off. Filters, when available, proved insufficient to keep out smoke. The apparent safety of many shelters and basements closed in by rubble was only temporary as the approaching fire increased heat and smoke. In others, detonations and explosions near by increased the pressure downward and directed the storm against the basements.

The Possibility of Escape.

Thus the picture changed from hour to hour. Whoever was still able to make his own decision had one of two alternatives: to stay or escape. Many looked into the streets, saw that everything was on fire, decided they could not get through, and withdrew into the corners of the shelters. Some tried to get out of the burning areas, and for them it was a race with death among explosions, fire bombs, machine guns, and falling flak. Besides all this, flames spurted through the streets and the wind caught up with many and threw them to the ground. There were screams from victims all around. No eyewitness mentioned screams with pain. Many people were caught in the fire. Many stated that the air "just didn't come any more" and breathing became very difficult. Otherwise they did not feel anything, and the rest went on over those who had fallen. One man was observed to fall. He was about to pull himself up with his hands when flames were seen to envelop his back and he burned within five minutes without changing his position.

The dead usually lay with their faces toward the ground. Many were lying in rows. Only a very few who had fallen got up by their own effort or with the help of others and reached safety in the areas which had not been hit. Some found safety in the bottom of a bomb crater; others found death by drowning in other water-filled craters.

Every possibility of escaping the firestorm behind rubble or remaining walls or corners was kept in mind. This was evident by the number of corpses found behind these ledges and corners. The same was true in open spaces where many sought safety behind tree stumps and parked cars.

The only safe refuge in all this time was the water of the canals and the port. Most of those who got there were entirely exhausted. Lips, mouth and throat were dry. They were blistered on the nose and ears, on the hands and face, and their eyes burned with pain and could hardly be opened after having been exposed to so much smoke. Many collapsed, then lost consciousness and died. Many jumped into the water. Even here the heat was hardly bearable. They took blankets and handkerchiefs, soaked them in the water, and then protected their heads and the uncovered parts of their bodies with the wet cloths. But the water evaporated so quickly that this procedure had to be repeated every few minutes.

It is striking that thirst was not a generalized symptom. Some victims could not take enough water, yet some in utmost danger of heat death denied a feeling of thirst. They did not seek water, although water was available, nor did they report that they sweated more than normally. Others, however, took off their sweat-soaked clothes as soon as they had reached areas safe from fire and excessive heat.

Only a few generalizations could be made from the remarks of those who came to safety. In the first hours after they had successfully escaped, some complained of headaches and slight drowsiness. The desire for sleep was present in all and sleep very deep. After awakening there were no sequelae.

Those Who Stayed

In the meantime the burned-out houses caved in. The rubble and debris on the streets prevented many from escaping. The heat decreased slowly, but the main danger was past. Many of the bodies were lying in the streets half clothed or nude.

The only covering that they always had on were their shoes. The victims' hair was often burned, but frequently preserved. A few hours after the start of the raid the corpses had a peculiar aspect; they seemed blown up, lying on their stomachs. The buttocks were enlarged and the male sex organs were swollen to the size of a child's head. Occasionally the skin was broken and indurated in many places and in the majority of cases was of a waxen color. The face was pale. This picture lasted only a few hours; after this time the bodies shrank to small objects, with hard brownish black skin and charring of different parts and frequently to ashes and complete disappearance.

At the same time fate had caught up with many of those in the shelters and the basements. In houses which had caved in through the effect of high explosives or fires, the bodies were found covered with rubble. The air raid tower shelters and also the larger number of the subterranean shelters withstood the explosions and fire. There was no doubt that in many a shelter, death had come to the occupants without any one ever suspecting it. Several persons were sometimes found sitting or lying in the most natural position; others were sitting in groups as if talking to each other and some had slipped to the floor from chairs or benches. The appearance of defense or escape movements could not be explained other than as death without premonition. In many shelters, however, bodies were found in a heap in front of the exit so that it must be concluded that escape was sometimes attempted.

In the shelters bodies assumed various aspects corresponding to the circumstances under which death had set in. Nowhere were bodies found naked or without clothing as they were in the streets. The clothes, however, often showed burned-out holes which exposed the skin. Bodies were frequently found lying in a thick, greasy black mass, which was without a doubt melted fat tissue. The fat coagulated on the floors as the temperature decreased. The head hair as a rule was unchanged or only slightly singed. The bodies were not bloated except for a few which were found floating in water which had seeped into the shelters from broken mains. All were shrunken so that the clothes appeared to be too large. Those bodies were called "incendiary-bomb-shrunken bodies" (Bombenbrandschrumpfleichen.) They were not always in one piece. Sleeves and trouser legs were frequently burned off and with them the limbs were burned to the bones. Frequently such bodies burned to a crisp weeks after death — apparently after oxygen had become available. In the same rooms with such bodies were found other more or less preserved or shrunken corpses and also some which had fallen to ashes and could hardly be recognized. Many basements contained only bits of ashes and in these cases the number of casualties could only be estimated.

Continual study was made of aerial photographs of target cities, before and after attack. In very general terms, here are some of the conclusions pertaining to fire spread in the target cities:

The factors that increased fire spread were: combustible buildings, combustible building contents, impossible fire fighting situations, limited water supply and, in some cities, development of a fire storm.

The factors that helped to limit fire spread included: parapeted fire walls, fire resistive exterior construction, separation between buildings and wide streets.

Fires in Japan[15]

In addition to the dropping of the atomic bomb on Hiroshima and Nagasaki, the Twentieth Air Force of the United States flew seventy-nine major missions over sixty-five target Japanese cities. This involved 13,365 aircraft sorties and a dropping of 93,000 tons of incendiaries and 650 tons of high explosive in fragmentation bombs causing nearly complete destruction of about 175 square miles, practically all in the centers of cities.

The primary objective of a number of these attacks was to start many fires in highly combustible areas in the shortest period of time. A single aircraft would drop 1,520 small oil bombs over an area of one third of a square mile. If only a fifth of these started fires there would be 300 fires, or one for every 30,000 square feet. The number of fires would increase in ratio to the increased number of planes dropping bombs.

The intent of this type of bombing was to create simultaneous fires that would blend together to form a thermal column and create a fire storm or at least a conflagration. In a saturation incendiary attack on Tokyo, March 10, 1945, the wind velocity increased from four miles per hour (at the beginning of the bombing) to hurricane proportions, causing great destruction. A total of 83,793 persons lost their lives.

On April 13 and 14 a similar raid over Tokyo occurred, this time with high explosives. The loss of life was relatively small although the entire city was practically in ruins.

In the saturation incendiary raid over Osaka (Japan's second city), four days following the first bombing over Tokyo, another near fire storm occurred. Almost sixty percent of the city was aflame very quickly, the communications systems were knocked out, water pressure in the mains dropped to near zero and fire fighting forces were practically surrounded by fire.

In a similar raid over Kobe 68,000 homes were destroyed and 242,466 people were left homeless.

The Atomic Bomb

At 8:15 in the morning of August 6, 1945, the first atomic bomb ever used as a weapon against a target exploded over the city of Hiroshima, Japan. Just forty-five minutes earlier an "all clear" had been sounded from a previous alert. Many people, workers, school children, employees, were out in the open and had not taken shelter. The bomb exploded slightly northwest of the center of the city, which was circular in shape with flat terrain. Practically the entire built-up portion of the city was leveled by the blast and swept by fire. A fire storm quickly developed and, two to three hours after the explosion, attained a velocity of thirty to forty miles per hour. Seventy to eighty thousand people were killed, missing, or presumed dead and then an equal number were injured.

At Nagasaki three days later the bomb was dropped from a B-29 at 11:02 a.m. No fire storm occurred, the area of nearly complete devastation was smaller, and between 35,000 and 40,000 people were killed, with about the same number injured. Without question, the bomb represented man's greatest and worst achievement in creating a weapon of destruction.

On August 10, 1945, Japan sued for peace and the world entered the atomic age.

On July 16, 1945, the world's first atomic bomb was detonated at Trinity site near Alamogordo, New Mexico. This was a weapon of twenty kiloton strength, the type that would be dropped on Hiroshima and Nagasaki in Japan, August 6 and 9, 1945. But this first blast signaled man's entry into the world of nuclear science and was a brilliant achievement to the scientists and technicians who had pooled their knowledge and talents in the urgency of wartime needs. Their reaction to that first atomic explosion was aptly described by one magazine, which reported: "They danced for joy." (Photo by Los Alamos Scientific Laboratory.)

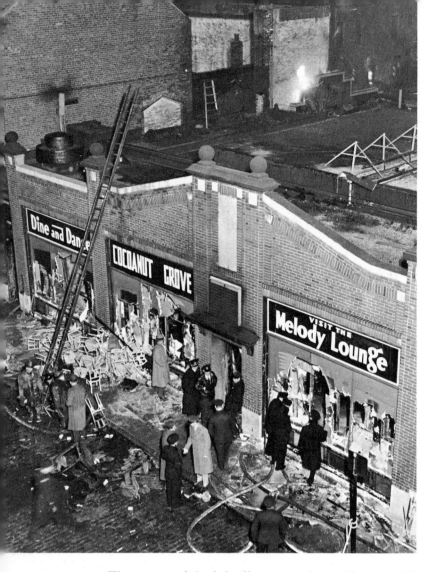

The worst nightclub disaster took the lives of 492 persons in the Cocoanut Grove in Boston, Massachusetts, November 28, 1942.

Panic in the Cocoanut Grove[16]

At the time when the United States was completely engaged in World War II a terrible nightclub fire occurred in Boston, Massachusetts. Unlike many other tragedies that were overshadowed by the war news, this incident made immediate impact on the national consciousness, and practical improvements in fire protection throughout the country were achieved shortly after the disaster. Among the important consequences was the knowledge gained in the use of blood plasma and the treatment of burns and general improvement in the enforcement of fire safety laws and ordinances throughout the U. S. and Canada.

The Cocoanut Grove was a typical nightclub, a one-story building with a basement cocktail lounge. Most of its exterior featured brick and stucco walls but inside the building there were low ceilings, combustible wall and ceiling finish, and flammable decorations. A holdover from prohibition days, the Cocoanut Grove was very popular and extreme congestion was common particularly on Saturday nights. November 28, 1942, was a Saturday toward the end of the football season, and the Cocoanut Grove was crowded to extremes. The official seating capacity of the building was something over 600, but on the night of the fire about 1,000 people were reported to have been in the building. The several bars were crowded, tables around the bars were filled to capacity and every available square foot of floor space was occupied just prior to the beginning of the floor show at 10:00 p.m.

Down in the Melody Lounge, a cocktail lounge in the basement, a sixteen-year-old bus boy was replacing a light bulb near an imitation palm tree. The exact cause of fire was never determined but it was agreed by witnesses that the point of origin was near this location. Within seconds, the fire flashed over highly combustible decorations, such as artificial cocoanut palms and cloth-covered ceilings and walls. It flashed up the stairs from this lounge cutting off the only visible means of exit. People on the main floor had no warning of the pending tragedy; some surviving witnesses said they first knew of the fire when a girl with blazing hair ran screaming across the room. Others first saw flames flashing through the air just below the ceilings. In seconds, throughout the building, there was a desperate rush for the exits.[17]

The main doorway, the only exit that most of those present knew, was blocked by a revolving door which quickly jammed. Some 200 victims piled up behind it. The flames also flashed through a corridor to the Broadway Cocktail Lounge and here 100 other victims piled up behind the door that opened inward, rather than outward to the outside doorway. Another door, leading to the street, was partially opened by an employee but other doors were locked. A few people escaped from the basement by crawling out of a cellar window, and some escaped through the small windows of toilet rooms. A few made their way upstairs to the second floor dressing room and escaped through windows onto roofs. But many people were quickly overcome by the noxious smoke and gases from the fire and collapsed at their tables without ever making a move toward the exits.

The Boston Fire Department was on hand immediately; in fact, just a short time before, four engine companies, two ladder companies, a rescue

One fact to remember: most people endangered by sudden fire within a building will try to escape through the same doorway they originally entered. If too many people rush for the same exit simultaneously, the situation can become disastrous. Whenever you enter any building take a few moments to learn the location of two or three possible exits . . . and make sure they are not locked or obstructed.

squad, water tower, a salvage company, the Deputy Chief of Division One, and the Chief of District Five had responded to a nearby automobile fire when cries for help, emanating from the nightclub, brought their immediate assistance. The first alarm of fire specifically for the nightclub was sent from a box alarm about 150 feet away from the Piedmont Street entrance. This was received at fire alarm headquarters at 10:21 p.m. and two more engine companies were dispatched, plus another deputy chief and a district chief. A third alarm came in at 10:24 p.m. and a fourth at 10:25. These brought fourteen engine companies, three ladder companies, and three district chiefs. A fifth alarm at 11:03 p.m. brought five more engine companies and two additional rescue squads were specially called.

Of necessity, the fire department had to carry out fire fighting and rescue operations simultaneously but, after the initial flashover, the fire was controlled promptly. The rescue operations were extremely difficult.

The work of emergency civilian medical services was credited with saving many lives. It was estimated that one Cocoanut Grove fire victim reached the Boston City Hospital every eleven seconds, a faster rate than casualties were taken to any hospital during London's first air raids. More blood plasma was used for burn and shock in Boston hospitals the first day of the disaster than was used in Hawaii after the Pearl Harbor raid. The Red Cross mobilized more than 500 workers within thirty minutes and the Nurses Aide Corp mobilized nearly 500 aides to work at the hospital, plus an additional 100 trained nurses. Two hundred and twenty-five units of dried blood plasma collected from volunteer donors were released to the Boston City Hospital and when the Boston Chapter of the Blood Donor Center issued a special appeal public response raised the week's collection of blood to an all time high of 3,789 units.

A disaster card system that had been developed for use in case of bombing attack on Boston was quickly utilized to list the victims of the tragedy because telephone calls soon were coming in from all over the country. One operator handled more than 1,000 calls within eight hours.

Four hundred officers and men from the First Naval District of the U. S. Coast Guard assisted in the fire fighting and rescue work.

The Cocoanut Grove fire received a tremendous amount of publicity and subsequent investigation brought criminal indictments against the building owners, the Boston Building Commissioner, members of the Police and Fire Department, a building designer, contractor and the foreman who participated in building the new bar and lounge.

The lessons of this fire were obvious and fortunately found their way into improvements of fire laws and ordinances throughout North America. Essentially they were these:

No place of public assembly should be filled beyond authorized seating capacity.

No combustible material should be used for decorations in places of public assembly.

Every building used for public assembly and every room and section thereof considered separately must have at least two means of exit located as remote from each other as practical.

If the building or room accommodates more than 600 people at least three exits must be provided; for more than 1,000 persons at least four exits.

Exits should be maintained free and nonobstructed at all times when the building is occupied. They shall be clearly marked and adequately lighted.

Exits must provide a clear path of travel to the street or to an open yard or court communicating with the street. Exit doors must swing with direction of exit travel.

Cocoanut Grove Floor Plan

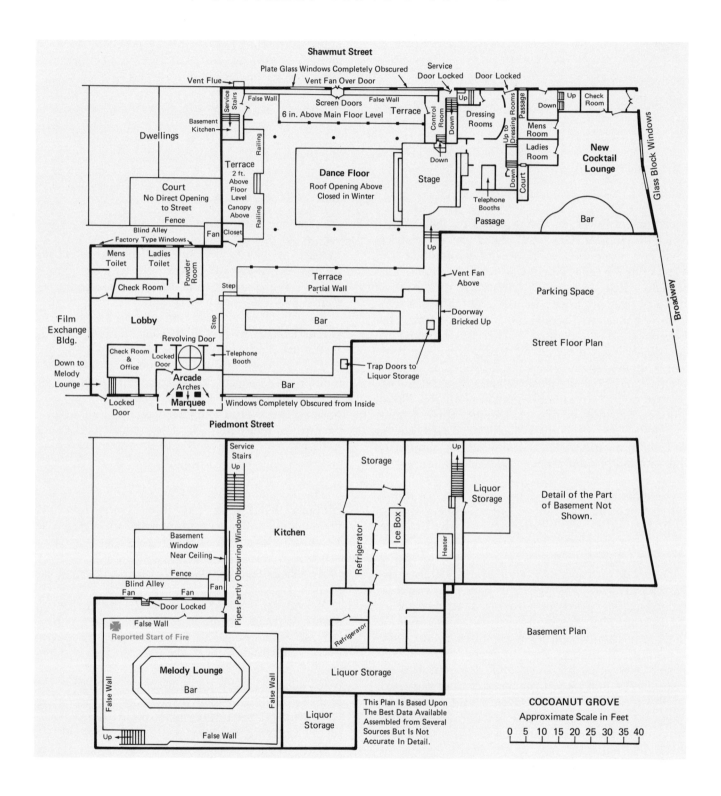

Shawmut Street

Plate Glass Windows Completely Obscured
Vent Fan Over Door

Vent Flue

Service Door Locked   Door Locked

Service Stairs

False Wall

Screen Doors
6 in. Above Main Floor Level   False Wall

Terrace

Up

Control Room

Dressing Rooms

Up to Dressing Rooms

Passage

Down   Up   Check Room

Basement Kitchen

Railing

Dwellings

Down

Mens Room

New Cocktail Lounge

Terrace
2 ft.
Above
Floor
Level
Canopy
Above

Dance Floor
Roof Opening Above
Closed in Winter

Stage

Down

Ladies Room

Court

Glass Block Windows

Court

Telephone Booths

Fence

Railing

Blind Alley
Factory Type Windows

Fan   Closet

Passage

Bar

Up

Mens Toilet   Ladies Toilet

Powder Room

Terrace
Partial Wall

Vent Fan Above

Check Room

Step

Parking Space

Bar

Step

Film Exchange Bldg.

Lobby

Doorway Bricked Up

Street Floor Plan

Revolving Door

Check Room & Office

Locked Door

Telephone Booth

Down to Melody Lounge

Arcade
Arches

Bar

Trap Doors to Liquor Storage

Locked Door

Marquee

Windows Completely Obscured from Inside

Piedmont Street

Service Stairs
Up

Storage

Up

Basement Window Near Ceiling

Kitchen

Liquor Storage

Detail of the Part of Basement Not Shown.

Pipes Partly Obscuring Window

Refrigerator

Ice Box

Heater

Fence

Blind Alley
Fan   Fan   Fan

Door Locked

Refrigerator

Basement Plan

False Wall

Reported Start of Fire

False Wall

Melody Lounge
Bar

False Wall

Liquor Storage

False Wall

Liquor Storage

This Plan Is Based Upon
The Best Data Available
Assembled from Several
Sources But Is Not
Accurate In Detail.

COCOANUT GROVE
Approximate Scale in Feet
0 5 10 15 20 25 30 35 40

Up

False Wall

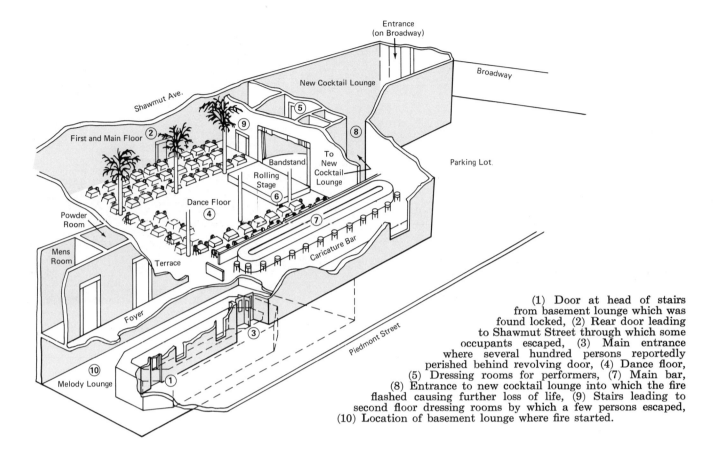

(1) Door at head of stairs from basement lounge which was found locked, (2) Rear door leading to Shawmut Street through which some occupants escaped, (3) Main entrance where several hundred persons reportedly perished behind revolving door, (4) Dance floor, (5) Dressing rooms for performers, (7) Main bar, (8) Entrance to new cocktail lounge into which the fire flashed causing further loss of life, (9) Stairs leading to second floor dressing rooms by which a few persons escaped, (10) Location of basement lounge where fire started.

WHAT TO DO IN A BURNING BUILDING [18]

1. Always be prepared for fire. When entering any building look around and choose your nearest exit. See what alternate path of escape may be available. Observe conditions and if you note locked or obstructed exits or any fire hazards, report them promptly to the appropriate authorities.

2. Remember that it is always dangerous to remain in a burning building. Fires often spread with incredible rapidity and cut off escape. Poisonous gases are likely to be generated in fires. Fire may be burning behind some partition and not appear serious but finally break out and involve the whole building in a few moments.

3. Take no chances of entering a burning building for the purpose of saving property. Only the saving of lives justifies taking a personal risk. Leave the job of fire fighting to trained firemen.

4. If there is a panic-rush for the main exit, keep out of the crowd and try to find some other means of escape. In some fires persons who have remained calmly in the building have been rescued unharmed after a panic-frenzied crowd has been crushed and killed in a jam at the main exit.

5. If forced to remain in a smoke-filled building, remember that the air is usually better near the floor. If you must make a dash through dense smoke or flame, hold your breath.

6. Remember that a temporary refuge may be secured behind any door. Even a thin wooden door will temporarily stop smoke and hot gases and may not burn through for several minutes.

7. Do not jump from upper story windows except as a last resort. Many lives have been lost in fires where people have jumped to their death even while firemen were in the act of bringing ladders to rescue them.

8. If burned in a fire report at once for medical treatment. Many burns which do not at first seem serious have fatal results. Inhalation of smoke and fire gases which may not at first seem serious can likewise cause fatalities, sometimes long after the fire.

Portion of the circus tent in Hartford, Connecticut, as men, women and children escape from the flames. One hundred and sixty-three persons died in the fire. (Acme Newspictures, Inc.)

Hartford Circus Disaster [19]

On the hot sunny afternoon of July 6, 1944, a large circus tent in Hartford, Connecticut, was suddenly enveloped by a flash fire. Within minutes, 163 persons died, sixty-three of them being children under the age of fifteen. More than 200 other persons were confined to hospitals with severe burns and fifty or sixty circus employees were treated by their own physicians. It was the worst fire tragedy ever experienced by "The Greatest Show on Earth."

The Ringling Brothers and Barnum and Bailey Combined Shows had presented circus performances throughout the United States for many years, under conditions that were substantially the same as those which resulted in the Hartford tragedy. The huge tent used for outdoor performances was 425 feet long, 180 feet wide and covered approximately 1½-acres. It was the usual type of large circus tent, supported by a number of heavy poles held by guide ropes secured to a double row of stakes approximately fifteen feet outside the tent. The largest poles had a maximum diameter of 12 inches.

The outer circumference of the tent contained a tier of stands having a seating capacity of 9,048 persons. This was distributed between 6,048 persons along the north and south sides of the tent, plus 3,000 general admission seats located at the two ends of the tent. For the Hartford performance, when the fire occurred, there were 6,789 paid admissions, indicating that approximately 7,000 patrons were in the tent. The circus had a staff of more than 1,300, an unknown amount of whom were in the tent when the fire started.

Just before the fire, wild animal acts had been completed in rings at the east and west ends of the tent. Temporary steel cages were placed in these rings for the animal acts and the animals entered from long temporary cage runways called "chutes" in circus terms. These extended to animal conveyance trucks located north of the main tent. One of these cage runways, extending across the main north aisle, served to block escape of many spectators after the fire started.

The fire began on or near the ground at the outside canvas immediately south of and about twenty feet from the main exit. Apparently it was caused by a discarded match or cigarette; whatever the cause, flames began spreading up the side of the tent. Someone apparently tossed three buckets of water on the flames without effect, because the fire increased rapidly and when it reached the edge of the top canvas the flame was about 2-feet wide at the point of contact. When a gust of wind occurred, the fire came into the underside of the tent and almost instantly the entry canvas was enveloped in flames. Ropes holding the supporting poles burned through almost at once and the large poles fell on the panic stricken crowd, causing several fatalities.

It was reported that most of the crowd were not disturbed immediately at the first sign of fire, apparently thinking it was part of the show or would be quickly controlled. But within seconds, panic developed. When some people ran down the aisle toward the east end of the tent they encountered the animal runway, but the steps leading over the runway proved entirely inadequate. Some tried to climb over the steel cage bars but could not. Most of the bodies of those who failed to reach the outside were piled four-deep against the cage obstruction in the main north aisle.

There were too many loose chairs in the seating area and, as frequently happens, people shoved the chairs and tossed them out of the way. Consequently, other persons were injured, or tumbled and fell across the chairs blocking their escape. Persons in the first five or six rows got out more easily. Those of the back section jumped ten or twelve feet to the ground and were able to escape under the back canvas. Most fatalities in this holocaust were due to severe burns, caused when the blazing canvas fell on the crowd igniting flimsy summer clothing. This contrasted with the fire experience at the Cocoanut Grove in Boston where many of the deaths were by suffocation. It is probable that some deaths were due to crushing by the panic stricken crowd, even though burns quickly followed.

Investigation after the fire disclosed a number of inadequacies which could have been avoided by proper fire protection. First, the practically new canvas of the tent had no flameproofing. It had been processed against water by the use of paraffin applied with gasoline as a solvent, a common waterproofing practice used by the circus.

There were nine means of exit from the tent but these proved inadequate, particularly the portion blocked by the runway chutes. One report stated

In a pathetic effort to be of some help the famous American clown Emmett Kelly carries a bucket of water, followed by another circus worker. As "Weary Willie," Kelly endeared himself to millions of adults and children, in audiences just like the one in this tragedy. At this writing he is still "wearing grease paint" at age 77; so is his son, Emmett Kelly, Jr., who copied his father's famous makeup. (Photo by Ralph Emerson, Stud Frontier Shop.)

that the circus would have required ninety-one units of exit but only forty-three units of exit actually existed. In addition to the animal chutes, the main exit aisle in front of the stand was not available to the audience during much of the performance because it was used by the performers. While the Grand Parade, chariot races and other activities were underway, spectators were confined behind the metal railing that circled the arena. There was only a narrow space between this railing and the front of the stand.

Of the many tragic stories and reports of human experience in this terrible disaster, one of the saddest concerned a small girl who died in the holocaust. No one ever came to claim her body.

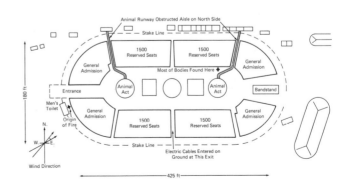

Bomber vs. Empire State Building[20]

View from upper floor of the Empire State Building showing opening made by impact of bomber. Debris can be seen on roof in lower center. Fourteen persons died in this accident. (Press Association, Inc., photo.)

On Saturday morning July 28, 1945, a B-25 two-engine medium bomber smashed into the Empire State Building in New York City creating a very dangerous and most unusual fire situation. The plane's pilot, in radio contact with LaGuardia Field disregarded advice to land there and started for the nearby Newark Airport. At the same time, he disregarded a Civil Aeronautics Administration Regulation that specified a 2,000-foot minimum altitude for aircraft over Manhattan. The morning was quite foggy but apparently the pilot had "contact visibility" for a distance of 2½-miles, which provided perhaps forty-five seconds apprehension of objects if the plane was at a cruising speed of 250 mph.

At 9:45 a.m. the bomber loomed out of the fog and crashed with great impact into the North wall of the Empire State Building, ripping a hole 18-feet by 20-feet in the exterior wall at the 78th and 79th floors. Flames shot as high as the observatory on the 86th floor, illuminating the tower by the glare. Part of the landing gear, another metal portion, and a radiator from the building smashed across the 78th floor, went through the south wall, then dropped to the roof of a twelve-story building across the street, starting a fire in a penthouse apartment.

Motors and parts of the landing gear from the plane crashed into an elevator shaft and fell to the sub-basement. Other sections of the fuselage were thrown as high as the 86th floor observatory. Flaming gasoline from the 1,400 gallon tanks splashed through the 78th and 79th floors, burning all combustible trim and killing or injuring all occupants of those floors. Burning fuel extended down stairwells to the 75th floor and people on upper floors were isolated by smoke and flame. Fortunately, because this was a Saturday morning, a diminished work force was in the building. Total casualties from the incident amounted to fourteen dead and twenty-six injured, some of whom had terrifying drops in elevators which fell from upper floors to the sub-basement when the aircraft snapped elevator cables.

The building had been designed to resist fire, and once fire fighters had climbed up the stairs from the sixtieth floor, which is as far as remaining elevators could take them, they were able to subdue the fire promptly. Four alarms brought twenty-three fire companies and more than forty pieces of apparatus. Flames on the 78th and 79th floors were controlled by handlines connected to the 8-inch building standpipes. More than 28,000 gallons of water was available for fire fighting.

The Fire Bell

". . . It rang box numbers to signal locations of alarms; a noon-time signal so people could set their watches and clocks; and then a 9 p.m. curfew when fire horses were bedded for the night and children were required to be home . . ." *History of the Newark, New Jersey Fire Department.*

The Curfew

There are still a few million parents and grandparents in America who can recall the sound of the evening curfew drifting over cities and towns in the twilight of a warm day. Quite a few communities retained this custom up till the start of World War II in 1941 — a time when many traditions ended.

The word "curfew" was derived from the French phrase "covrefeu", meaning "cover fire". In European countries it was the practice to place special covers over fires on the hearth to minimize the chances of sparks igniting nearby combustibles and causing a dangerous fire. The ringing of the curfew bell also reminded the people that it was time to retire to bed, because watches and clocks had not been invented then. The curfew custom (a deep-sounding whistle blast, or tolling of a bell) continued in America and was especially useful for testing the means of sounding a fire alarm. It is no longer effective for getting young persons home by 9 o'clock at night!

In Perth Amboy, New Jersey, June 23, 1949, fire started in an oil refinery. At one moment during the blaze this 2,000-barrel asphalt tank was projected upward like a rocket but fortunately fell without harm. This photo emphasizes the result when the top of a vertical tank is stronger than the bottom. Thus, if the bottom fails under pressure, there is a rocket effect. If the tank is designed so that the top is weaker than the other portions, fire and pressure will be relieved through the top, usually without serious consequences. (Wide World photo.)

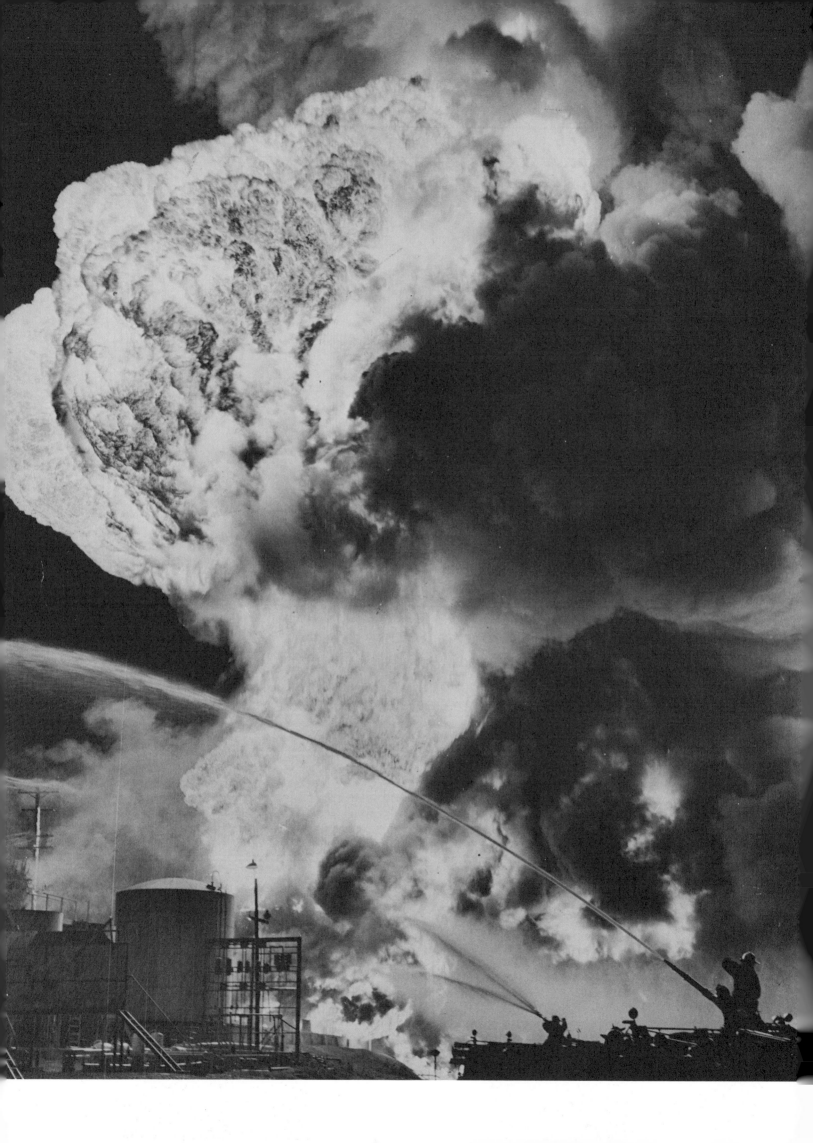

(Left) On July 5, 1961, this fire occurred at a petroleum bulk plant in Richmond, Virginia, and caused the death of one man and monetary loss exceeding two hundred and sixty thousand dollars. Shortly after 1 a.m., a barge was unloading 450,000 gallons of gasoline into the shoreside tanks of the plant. About forty-five minutes later a heavy concentration of vapors was ignited by an undetermined source. The flash spread from the barge to several dwellings across the street. Four men on a tug tied to the barge leaped overboard and swam to the opposite bank of the river. The clothing of one man ignited and he ran for fifty yards before someone tackled him and ripped away the burning cloth. With burns over ninety percent of his body, he died several days later.

(Photo by Richmond Times Dispatch.)

1946-1960

Chapter VII

When World War II ended in 1945, the people of the United States and Canada were prepared for a return to normalcy, but no one could have predicted the tremendous changes that soon would occur. Both countries had undergone severe hardship and sacrifice during the ten years of depression and the four years of commitment to a total world war. In the latter years there had been rationing of basic essentials — food, clothing, gasoline, automobiles, and most other materials. Building lights and street lights had been dimmed or completely shut off during nighttime hours because of potential bombing raids. Most citizens, who were not engaged in military service, were working long hours, day and night, to meet demands of the war effort.

After such sacrifice and restriction it is no wonder that explosive growth and startling technological change began almost before the troops started home. And because of the changes, new fire problems emerged and fire protection technology entered a whole new era of expansion. Consider these facts:

• Research for the wartime effort had uncovered many new processes and materials that would have application in the civilian market as soon as security restrictions were lifted.

• A large portion of population in both countries had moved during the depression and wartime years and were ready to move once more to cities and towns that offered the greatest promise of economic gain and security.

• Frequency modulation of radio, greatly improved for military operations during the war, would be released to public service agencies and

fire department alarm and response would improve tremendously.

• Every city and town would be affected by the tremendous building programs that brought construction of millions of new dwellings, apartment buildings, stores, churches, industrial plants, and other occupancies; at the same time older, inefficient buildings were doomed, and were to become major fire problems.

• In the United States a huge new highway system would be constructed that would stimulate the building of new communities, would speed transportation, would increase the general mobility of population, and simultaneously would create very serious transportation fire and explosion problems.

• The need for providing basic fuels for the new technological society would increase the shipment of gasoline, oil, natural and propane gas, plus thousands of new chemicals and derivatives, each of which had inherently hazardous characteristics.

• Many new synthetic materials would be introduced, with beneficial and problematical results.

In those postwar years, many significant fire incidents brought their individual lessons and paradoxically led to major improvement in fire safety. The principal occupancies affected were: hotels, hospitals, nursing homes, schools, oil refineries, aircraft and industrial plants. At the same time, the methods of fire protection, fire prevention and fire control were improving and subsequently would help to diminish the frequency of such disasters.

189

Open stairways and smoke-filled corridors forced some LaSalle Hotel guests to use exterior fire escapes. Three interior enclosed stairways did not discharge to the outside at the ground floor.   (Acme Photo.)

A "Fire-Resistive" Hotel[1]

This was the first major post-war hotel disaster that caused extensive loss of life. The basic structure was built of fire-resistive materials but combustible interior finishes created tremendous flames.

By 1946 millions of people had traveled sufficiently to appreciate the comfort and convenience of a modern hotel. So, when the La Salle Hotel in Chicago, Illinois, burned on June 5 with a loss of sixty-one lives the public was shocked by the tragedy, because it could readily identify with the type of building. The 1,000-room hotel was filled to capacity at the time of the blaze but fortunately many of the occupants were able to escape.

The fire was observed at three points near the first floor cocktail lounge and elevator shaft just before an alarm was telephoned to the fire department at 12:35 a.m. First responding companies found the entire lobby roaring with flame; subsequently it was estimated that the fire had been burning twenty to thirty minutes before the alarm was given. The flames flashed quickly through the lounge and coffee shop and into the hotel lobby which had large quantities of walnut veneer panel-

ing and other combustible interior finish. The fire quickly spread up and over stair shafts to corridors on the third, fourth, and fifth floors, trapping many occupants. Much loss of life occurred in the third floor corridor as hotel guests tried to escape from their rooms. If doors and transoms were left open, rooms on the third and fourth floor were burned out.

Many people fell to death or tried to escape from windows by climbing down bed sheets tied together. Others dropped to their fate, perhaps fifteen or twenty; most might have been saved if they had waited for rescue. The Chicago Fire Department dispatched six aerial ladder companies to the fire and fire fighters raised hundreds of feet of ladders to rescue occupants of the six lower floors who were most seriously endangered. Hundreds of other people were able to find the path to exits and escaped by descending fire escapes to the ground.

This was one of many fires that magnified the hazards of combustible interior finish, but it was a long time before state and local fire safety regulations were changed to minimize this hazard.

The La Salle Hotel fire, and other fires that occurred within the following ten years, emphasized certain fundamentals of major significance. First was the behavior of people when they were faced with an emergency of tremendous danger and overwhelming possibility of tragedy. In some respects, their behavior did not vary from the behavior of victims of the Iroquois Theatre disaster, the Natchez Dance Hall, and Cocoanut Grove victims, but the decision of the La Salle guests to leap to death rather than to take a chance against the fire potential and wait for rescue was a tragic indication of panic and despair.

Second, the use of combustible fiber board, which in those years was not much more than untreated pressed paper, was a major contribution to fire spread, and to almost explosive development of heat and flame.

Laminated plywood was also used in many buildings, with no regard for its fire potential. Studies of this material after the La Salle Hotel fire proved its potential destructiveness.

Then there were the open stairways and unprotected elevator shafts which existed in so many old hotels and apartment buildings. As in the La Salle Hotel, these openings virtually assured the deaths of people on the upper floors.

This major hotel catastrophe focused attention on certain weaknesses that probably existed in many hotels and motels of that time. For example, above the cocktail lounge on the first floor there was a suspended combustible ceiling made of such material that fire could develop rapidly and produce intense heat and toxic gases.

• The fire produced additional evidence that people in an emergency would react in an extreme manner, possibly leading to their own deliberate suicides or accidental death.

• The La Salle Hotel, among many others of that time, was thought to be of "fire proof" construction. Today, we acknowledge that nothing is "fire proof"! Every material can be affected by fire.

The La Salle Hotel was twenty-two stories in height, with a basement and sub-basement, and was constructed of protected steel frame and reinforced concrete. Exterior walls were of the brick panel type, with interior partitions of 3-inch hollow tile, plastered on both sides. However, all the bedrooms featured wood panel doors and wood transoms, some of which were open at the time of the fire. Most of the window sash was of wood.

• There were three enclosed, noncombustible stairways but they did not open directly to the outside at ground floor level. At the time of the fire one door leading to the enclosed stairway near service elevators on the mezzanine floor had been left open. This stairway, which offered the quickest exit to the outside on the first floor, could not be used.

• Apparently nobody realized the inherent danger of the beautiful walnut veneer paneling, so prominent in the large lobby and mezzanine. Nor was anybody aware of the unprotected concealed spaces in the lounge and coffee shop and the vent openings into the elevator shaft. The cocktail lounge, where fire originated, included gypsum block filler placed on top of wood studs and covered with an attractive but highly combustible interior finish.

• Perhaps the most important building deficiency was the placement of exhaust ventilation from the cocktail lounge to discharge into the masonry elevator shaft. As fire developed, this ventilation permitted rapid extension of heat and fire gases to upper floors.

• There were two possible sources of ignition: overheating of a lighting fixture in the cocktail lounge, or, a carelessly discarded cigarette in the elevator shaft. Either could have started this fire.

Fire fighters rescuing hotel victims on fire escape. (International News Photo.)

Fire fighters on aerial ladder rescue woman trapped on fifth floor of the Hotel Canfield. Thirty-seven persons were rescued by ladders from the two hotel sections. (Associated Press photo.)

Canfield Hotel[2]

It is difficult to believe that owners and managers of hotels, fire department inspectors, and city officials could not anticipate the potential destructiveness of fires in hotel buildings immediately after World War II. Most of these buildings had existed for thirty, forty, fifty years, or more and obviously were deficient in convenience, heating, air conditioning, spatial arrangement, and overall fire protection and safety. It was easier, and supposedly more economical, to allow such buildings to function until a major problem occurred.

The Hotel Canfield experienced that major problem on the night of June 9, 1946, when fire took the lives of fourteen men and five women and caused serious injuries to twenty other guests. The responding Dubuque Fire Department had little chance to control the fire situation because their efforts had to be directed to rescuing the surviving victims. The unsprinklered, ordinary brickjoisted portion of the hotel included an open wood stairway up to the three top floors of the building. When the six-story fire-resistive annex was built in 1925, an interior, noncombustible stairwell was constructed serving each floor with automatic fire doors leading to the stairwell at various floor levels. These doors were designed to close automatically and became quite important as the fire intensified.

Access to the fifth and sixth floors of the fire-resistive section was from elevator landings in a frame, iron-clad passage, built above the roof of the four-story combustible section.

There were other facts before the tragedy:
- Following the disaster of the Cocoanut Grove fire in Boston, the Dubuque building official had required that additional exits to the outside be provided from the lounge on the first floor.
- It was the practice of the bartender to require waitresses to collect cigarette butts and paper napkins left on table tops and to deposit such refuse in a paper container in a small closet on the street side of the bar. This might have been a source of ignition.
- Once the fire started it spread rapidly out of the closet to the highly combustible fiberboard interior finish of the bar room. This fiberboard was glued to ⅜-inch plasterboard, which was nailed to suspended 2- by 4-inch stringers. The concealed spaces above the suspended ceiling were not protected by automatic sprinklers or automatic fire detectors and alarms.

The substitute night clerk on duty reported that he first noticed fire emanating from the unprotected door opening between the lounge and the hotel lobby. The hotel manager attempted to warn guests on upper floors and a telephone alarm was sent to fire department headquarters at 12:39 a.m. (This was fifteen minutes or more after the fire was discovered!)

When first fire companies arrived, flames were bursting prominently from third, fourth, and fifth floor rooms. The combination of highly combustible fiberboard interior finish and the open combustible stairway that led to the top of the building, proved to be disastrous.

Despite the sudden calamity of this incident, fire fighters and civilians used life nets to save twenty-seven persons. Aerial ladders and ground ladders were placed to rescue thirty-five of the one hundred and twenty-nine permanent and overnight guests. Many persons escaped by using exterior fire escapes at the rear of the adjoining fire-resistive section, but another exterior fire escape could not be used because flames broke through a building wall or window opening to create a serious threat.

Following investigation into this fire came these conclusions of the investigating group:

• It was found that lives had been lost in the combustible *and* the fire-resistive sections of the hotel.

• Fire raging up the open stairway in the old building mushroomed, driving heat and smoke into corridors of the fire-resistive section.

• A fire door was wedged open with luggage, which indicated that one victim had tried to get out of the building the way she came in, without knowing there was a fire escape safe to use less than seventy-five feet away from her position.

• Doors to rooms off corridors in the fire-resistive section were wood paneled, with metal ventilating grills. In outside rooms where doors and transoms were closed there was little evidence of smoke and fire damage.

• Subsequent investigation indicated that if all guests had been informed on checking into their rooms that the building offered two safe means of exit, aside from the open stairways and the unsafe fire escape, the loss of life and the amount of injuries would have been far less in this disaster.

In 1959 Alaska and Hawaii became the newest of the United States of America.

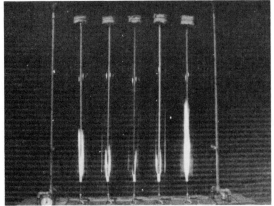

35 seconds

60 seconds

90 seconds

120 seconds

Burning Characteristics of Building Materials[3]

After the tragic fire in the La Salle Hotel in Chicago, a great deal of attention was given to the flammability of wall covering and other materials inside buildings. Some panelling from the hotel was brought to Underwriters' Laboratories, Inc., in Chicago and tested to compare its burning characteristics with other known material. The late A. J. Steiner of the laboratories developed a practical and dramatic test showing how laminated wood and other flammable wood particles react when exposed to ignition temperature. Here is a sequence of pictures illustrating how different types of wood burn under controlled laboratory conditions. Careful analysis of these burning patterns leads to objective conclusions which, in turn, can strengthen regulations against materials that are most hazardous to fire safety.

In the eight photographs shown on these pages the five samples of wood are exposed to the same ignition temperature and progressive destruction by fire is obvious. From left to right, they are: (1) A piece of wood dried to a moisture content of seven to ten percent, similar to what would be found in a building during a heating season.

(2) This piece had about the same amount of moisture that would be contained in wood in a building, or in an industrial establishment, under high humidity conditions.

150 seconds

180 seconds

210 seconds

(3) This piece has been treated with a chemical throughout its cross section with the intent of keeping the wood material from spreading the fire.

(4) This piece was painted with a so-called "fireproof" paint or coating.

(5) This piece is the same as (1) except that it is laminated, similar to the wood material in the La Salle Hotel. Delamination occurred under heat. In fact, it is obvious that this material burned most vigorously; even so, the flame spread in (1) and (2) does not give much comfort!

In the UL tests all material were conditioned to a moisture content stability under standardized limits of approximately seventy degrees Fahrenheit temperature and thirty-five percent relative humidity. Moisture content and air movement were controlled, together with the means of ignition, to permit the closest duplication of conditions.

Subsequent to these tests, the same type of analysis was given to combustible acoustical tile and other interior finishes prevalent during the 1940s, 50s and 60s. (*Photos by Underwriters' Laboratories, Inc.*)

240 seconds

Exterior view of burning Winecoff Hotel in Atlanta, Georgia.

The Winecoff Hotel Disaster[4]

The third major hotel fire in 1946 occurred in the Winecoff Hotel in Atlanta, Georgia, on December 7. At least 122 persons died in the fire, or by jumping or falling from the building. Severe heat developed in the fifteen-story building as the fire spread from the ground floor up through a single, narrow (3-foot, 5-inch), open stairway that served all the upper floors, and acted like a chimney to funnel the heat. The stairway was not the cut-off, smoke-proof, exit tower usually required for such buildings; it did not even have doors at each floor.

One of the important factors in the spread of heat and fire gases was the fact that many guest rooms had open transoms which admitted the smoke, heat and flame. These fire products not only threatened the life safety of the guests but also served to involve combustible furnishings

In 1946, most hotels had wooden doors and transoms and combustible furnishings in each room. Many of these buildings had an open stairway that extended to the top, forming a tunnel for the spread of heat and fire gases.

within each room. In general, from the third to the sixth floors of the building, where the transoms were closed, the wooden doors to the rooms held until fire fighters advancing lines up and down the narrow stairway could knock down the fire. The occupants of these lower floors also were within reach of fire department aerial ladders; above the sixth floor, they had to find some other means of exit. At least one couple had the presence of mind to close the transom, cover the door with a mattress and other bed clothes, and toss water over these items to cut down the heat seeping into the room. Then they went to the window and opened it slightly top and bottom. They crouched on the floor, breathing the relatively cool, clear air that came in through the bottom window opening; thus they were able to survive the extreme fire condition.

The most horrifying aspect of this tragedy was the number of persons who chose to jump from windows rather than stay and try to survive the flames. Several fire fighters attempting rescue were injured when struck by some of these panicked victims. Fire department ladders were broken by falling bodies and the new 100-foot aluminum ladder of the Atlanta Fire Department was severely bent by impact of one body, although it remained in service.

Fire fighters were able to get above the fire and knock it down before it spread to the top stories of the hotel. They used hoselines from standpipes in buildings across a ten-foot alley and a fifty-eight-foot wide street, as well as in the hotel.

The exact cause of the fire was not determined. A matteress in the third floor corridor might have been ignited by a cigarette or possibly started burning in a guest room and had been dragged out to the corridor. Whatever the cause, at about 3:15 a.m., a bellboy, responding to a call on the fifth floor, found the stairwell and hallway well involved by fire. The initial alarm to the fire department was at 3:42 a.m., a considerable delay after fire was discovered. The hotel did not have any automatic alarm or automatic sprinkler system.

Hotel Fires

Like other seaside resort areas, Atlantic City, New Jersey, has a large number of hotels and rooming houses built closely together. Now and then fire involves one or more of these buildings and difficult fire control problems develop. The photo above shows part of three hotels and nine rooming houses that were swept by fire January 7, 1952. Scenes like this have been repeated many times in different resort areas.

In the off-season, such hotels and rooming houses are vulnerable to arson, in addition to their normal susceptibility to fire. During the tourist season, and especially in hotels of year-round occupancy, management must be deeply concerned about the potential of fire. Its principal responsibility is protecting the lives and safety of the hotel occupants.

Hotels usually have open stairways; long, undivided hallways; large convention rooms, or ballrooms; one or more restaurant and dining areas; cocktail lounges; and large storage and service facilities, usually in the basement. Each of these areas can be susceptible to ignition and, if the hotel has not been adequately designed and protected, extensive fire may develop.

Smoking materials and matches have always been a frequent cause of hotel fires. Faulty electrical equipment and restaurant cooking facilities also have been sources of ignition. Fires of suspicious origin in year-round hotels amounted to thirteen percent of the incidents in one hotel study; in seasonal hotels this figure was thirty-one percent.

If fire does start, it is essential to inform *all* occupants quickly, then move them out of the building efficiently and safely. Automatic sprinklers provide the best of protection but they must be supplemented by a good alarm system; enclosed stairways, elevator shafts and vertical openings; adequate exits; good emergency lighting that comes on automatically when the main power fails; and standpipe hose, located in the right positions with adequate water supply.

In older hotels, open transoms above doors have been important in allowing fire to spread to combustibles in room interiors. Combustible ceilings and wall materials also have been major factors in fire spread.

Air conditioning systems may serve to move smoke, heat and other products of fire unless automatic dampers and other functional equipment operate quickly to control the spread of smoke, heat and toxic gases. But the main task is to move all occupants out of the building before the fire becomes large enough to hinder escape. This requires careful preparation of the hotel staff, long before a fire emergency. People who stay at hotels should learn at least two ways of exit from each floor and should take time to read the hotel safety regulations, usually posted in each room. The most consistent warning is, "Don't smoke in bed!"

Texas City Disaster[5]

The worst explosion in the history of the United States occurred in Texas City, Texas, April 16, 1947. In addition to causing tremendous destruction of property it took the lives of all but one of the Texas City fire fighters and destroyed their four pieces of fire apparatus. Four hundred and sixty-eight people died in the blast and subsequent fire devastation and property damage was estimated at fifty million dollars.

The blast resulted from fire and explosion involving two ships docked in the Texas City waterfront, both carrying large cargoes of ammonium nitrate. For many years this chemical has been used as a commercial fertilizer, but it is an oxidizing agent that supports combustion and is capable of detonating when heated under confinement that permits a high pressure build-up. It is also sensitive to strong shocks, such as from an explosive. It is especially dangerous to permit contamination of ammonium nitrate with oil, charcoal, or other combustibles and sensitizing substances because the entire combination may explode. This was the primary lesson of the Texas City disaster.

A contemporary report of this unusual incident contained the following statements:

"The force of the blasts was unparalleled, considering the nature of the material, its fire exposure and its storage. Previous ammonium nitrate explosions had been under vastly different conditions, occurring where an initial detonating agent had been present, or during the manufacturing processes where chemical stability had not been secured. Previous shipboard fires had not resulted in explosions.

"The violence of the Texas City explosions may be attributed to the chain reaction initiated by the chemical decomposition of the nitrate, the presence of impurities, its density of stowage, the fine granular nature of the material (treated to prevent caking), the breakage of the paper bags permitting further contact with foreign material, and the delay in application of large capacity hose streams. Confinement of the cargo in the ship's hold presented an additional hazard to life and property as the explosions shattered the vessels so totally that no portion remained visible above the water level. Flying metal weighing many tons and piercing fragments resembling shrapnel fell throughout the area, killing and maiming, piercing structures, oil tanks and automobiles."

This remarkable incident began about 8:00 a m. on the morning of April 16 when longshoremen beginning the day's loading operations discovered fire in the No. 4 hold of the *S. S. Grandcamp*, a former Liberty ship which had a cargo of ammonium nitrate. The longshoremen and some of the ship's crew tried to extinguish the fire with small quantities of water, then they tried using steam.

The Texas City Fire Department, which was all volunteer except for a paid chief, responded to a belated telephone alarm and was using hose streams on the cargo when, at 9:15 a.m., a tremendous explosion shattered the area, killed twenty-six of the twenty-seven fire fighters on duty and completely destroyed their four fire trucks.

It was not until the following morning at about 1:15 that the second explosion occurred aboard the S. S. High Flyer, docked in a separate slip from the Grandcamp. This had become heavily involved in fire, probably because of exposure to the Grandcamp explosion. Flames were first detected aboard the High Flyer in midafternoon but no effort was made to control the flames because all disaster crews were working in rescue operations. There was no prearranged mutual aid but fire departments in the surrounding area including the Houston Fire Department responded on their own initiative.

The incredible power of the explosive waves following the two principal explosions is indicated by the collapse of two fire-resistive, reinforced concrete piers. These buildings, one of which was two stories in height, were reduced to rubble. The noncombustible pier structures were merely twisted masses of metal in which fire burned for days.

Nearby buildings of a major chemical plant were shattered by the explosion; so were other buildings in the industrial area. Refinery storage tanks ignited and burned continuously the day following the explosion, because no fire fighting could be directed to their control.

Ammonium nitrate had not been considered a hazardous commodity but the crew of the Grandcamp knew that the fertilizer was highly soluble in water. Water applied to a fire involving ammonium nitrate might cause minor explosions which would scatter burning materials due to the sudden formation of steam but fire fighters could get some protection behind a barrier at appropriate distance. Most dangerous is the explosion potential of this heated fertilizer compound.

Monsanto chemical plant is in foreground. Arrow points to where the Grandcamp was berthed. In background two tanks are burning on property of Texas City Terminal Railway Company. (Acme photo.)

Fire fighters and dock workers hauling hose to combat the fire in the Grandcamp just before the first explosion. Twenty-seven were killed just after this picture was taken. (Press Association photo.)

Closeup of uncontrolled fire in one of the refinery tanks. (Los Angeles Fire Department photo.)

Tragedy at St. Anthony's Hospital[6]

Just before midnight, April 4, 1949, a rapid-spreading fire started in St. Anthony's Hospital in Effingham, Illinois. Within minutes it took the lives of seventy-four patients and injured many other persons. Among the fatalities were eleven newly born infants. It was a tragic night of terror, sorrow and frustration for everyone at the scene. However, like other major fires, this one made a deep impression on the public, the medical profession and hospital administrators. In due course, many hospitals across the nation received fire protection improvements as a consequence of the lessons of this fire.

At the time the fire occurred, St. Anthony's was an old hospital, spotlessly clean, with excellent housekeeping, but with few features planned for limiting fire spread. It was a 100-bed hospital, but had more than 100 patients the night of the fire; altogether there were 128 persons in the building, including ten members of the hospital staff and two bedside visitors. The infants were in a small nursery on the second floor.

The original 2½-story and basement section of the hospital had been built in 1876, principally of brick and timber. There had been three open wood stairways from basement to attic, without fire doors, or other partitions or protection. At some

early period one of these stairways had been torn down and the open space had been covered with flooring. The other two stairways remained in service. Interior finish of the building was wood lath and plaster, and the doors and trim also were of wood. On the third floor, room separations had been constructed by using cellulose fiberboard to cover wood-stud partitions.

About 1912 or 1913 another three-story section with full basement was added. This had brick walls, wood joist floors, a flat deck roof above a concealed space, and wood lath and plaster on the interior. It also had an open wood stairway without fire doors, and a combustible laundry chute from the basement up to the third floor. The corridors of the original building were extended into this new section. In the basement were the laundry and maintenance shop.

In 1943 ceilings of the open corridors, except in the basement, were soundproofed by application of combustible acoustical tile which extended down the walls eighteen inches from the ceiling. The rest of the side walls were covered with a type of oil cloth material. There were no planned barriers to fire spread in the hospital. As in many other old buildings of that time, the hospital had specific deficiencies that made it ready for disaster: a large amount of combustible interior material, open corridors, stairways and vertical shafts, no sprinklers, no fire detection units and no alarm system, for local warning, or for immediate, direct contact with the fire department. It was operated by the Sisters of St. Francis who resided in a convent connected directly to the hospital by a passageway into the basement.

The cause and precise location of fire origin were never determined, but one of the staff, Sister Eustasia, smelled smoke on the third floor of the east wing and telephoned Sister Anastasia, the night superintendent, who was at the switchboard on the first floor. Sister Anastasia immediately called the hospital's chief engineer, her Sister Superior in the adjoining convent building, and the Effingham Fire Department. There were no time recording devices at fire headquarters but an Assistant Chief stated that the alarm was received at 11:48 p.m.

The chief engineer raced from his home about one hundred yards away and tried to fight the fire with a portable extinguisher directed into the bottom of a laundry chute. (After the fire, four emptied 2½-gallon soda and acid extinguishers were found in the rubble near this location.) Several nuns tried to assist him but were driven out by smoke and heat. It is estimated that twenty-two members of the Effingham Fire Department arrived on hospital grounds within ten minutes after receipt of the alarm but by that time fire had broken through the roof, the third floor was completely involved and most of the victims were dead. In response to a radio alarm, fire departments responded on mutual aid from distances up to sixty-two miles.

Emergency evacuation of all patients in the hospital was impossible. Forty-two persons died on the third floor, twenty-nine on the second floor and three on the first floor. The combustibility of the structure and the rapid spread of fire on the surface of the corridor interior finish, together with the open stairways, prevented any use of two exterior fire escapes and two slide escapes. Patients jumped from windows to injury or death, even though neighbors, nuns, nurses and others responded to the scene, spread mattresses and helped move ladders to the windows.

The heroism of survivors is ofttimes overlooked in the heroism of the dead, but Fern Riley, young nurse in charge of the second floor nursery, in the early stages of the fire had only to step outside the window of the nursery to the fire escape and safety, but died with her small charges facing certain death. Hundreds of proud fathers must have peered through the plate glass windows at the end of the south corridor at their wrinkly new-born. The same window must have held terror to Fern Riley as she must have seen smoke and flames racing toward her through the 120-foot long open corridor and spreading down the open stairway from the third floor, just outside.[7]

To other hospital administrators the St. Anthony disaster is a tragic invitation to consider their own situation, and to follow qualified advice and assistance in the evaluation of the fire hazards, if any, present in their hospitals. A group evaluation should be made on the supposition that destructive fire may originate in any room or location within a building whether in a visible or concealed space, and in the light of available facilities for the detection and extinguishment of fire originating from any possible cause.[8]

1

2

3

4

Test of An A-Bomb[9]

On March 17, 1953 the U. S. Atomic Energy Commission tested an atomic bomb to measure its heat and blast effects on a typical dwelling. The building was placed at 3,500 feet from where the bomb was activated on a tower. These pictures show an exceptional sequence of destruction.

This bomb was slightly smaller than the type exploded over Hiroshima and Nagasaki in Japan in World War II. Those bombs were of twenty kiloton strength; this device was of the fifteen kiloton class.

An atomic explosion releases energy as heat, light and nuclear radiation. The heat energy which is released instantaneously produces very hot gases at a high pressure. The outward movement of these gases creates a shock wave capable of severe destructive effects. The radioactive fission products are released a few thousandths of a second later.

Prior to the test it had been calculated that no fire would occur on the exterior of the test house at the 3,500 foot distance. Picture Two shows smoke and charring before dissolution of the roof and walls begin in Picture Three. Then the building shatters under the tremendous force of the blast.

At the end of one second in a blast of this capacity, the fireball reaches its maximum radius of 450 feet and begins to rise like a gas balloon. The shock front of the air blast is visible 600 feet ahead of the fireball. By the end of ten seconds the intense luminosity of the fireball has almost died out but the shock wave has traveled 12,000 feet (more than two miles) and passed the region of maximum damage, then formation of the typical atomic cloud begins.

Most of the blast damage from an explosion of this size occurs within a radius of 12,000 feet during the first ten seconds after detonation. By thirty seconds, at a distance of about seven miles, almost all the immediate energy has been dissipated and only light damage results, such as to glass and windows in a building. At seven to eight miles, an observer may feel a jolt or strong push from the air blast and hear one or more loud slaps of sound. The blast waves from one detonation were registered on equipment and heard distinctly more than six hundred miles away from the test site.

In the photos on these pages, reading from 1 to 8, the first shows the test house 3,500 feet away from the bomb tower in the first 1,1000th of a second after the bomb was exploded. In photo 2, at 11/24ths of a second, heat of the fireball has ignited the thinnest wood in shutters and door panels, and the wood walls facing the blast are scorched. Whitewash instead of paint was used on the house exterior to minimize ignition.

The period of heat radiation was very short, but the quantity of heat per square inch of wood surface was very large, which caused charring and smoking, but no discernible flame.

In picture 3, at ¾ths of a second, the charring and smoking are snuffed out as the first blast effects hit. Then, at 19/24 second, the stronger effect of the blast waves reaches the house, as in picture 4. The severe air movement begins to tear the house apart in the sequence in pictures 5 to 8. Where surfaces of the building show as white, this is a reflection of the fireball from the unburned boards and shingles. The charred wood shows as black.

No sustained fire resulted in this building after the blast, although it was demolished. Stoves, electric wiring and other possible sources of secondary fires had been removed before the test, so, in that respect, it was not a "typical" dwelling.

5

6

7

8

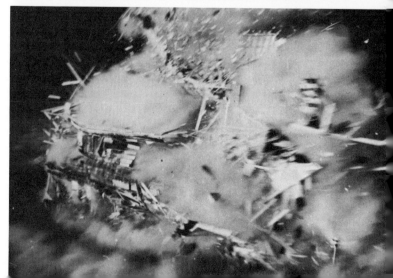

Air view of Whiting Refinery fire. Note how small fire apparatus and hose streams appear against this tremendous fire. (Wide World photo.)

An explosion, a boilover and hasty emergency diking were interesting consequences of this refinery fire that lasted for eight days before final extinguishment was completed.

Explosion and Boilover[10]

Shortly after six o'clock on the morning of August 27, 1955, a violent explosion occurred in a hydroformer in the refinery of Standard Oil Company (Indiana) in Whiting. The rapid spread and behavior of the resultant fire presented an unusual fire control challenge.

A hydroformer is used in the process of making high octane gasoline and other products. At this refinery, the hydroformer was in a steel tower that was nearly 265 feet high. When the explosion occurred this tower was completely shattered and huge fragments were sent flying for several hundred feet. A three-year-old child was killed by a flying piece of steel which penetrated the roof of his home. Two houses were completely destroyed, eighty received major damage and a hundred were damaged to a lesser extent. Fragments also penetrated nearby storage tanks permitting different grades of flammable liquids, including crude oil and aviation gasoline, to flow freely on the ground. Fire started almost immediately and it was estimated that one million barrels of flammable liquids burned in the subsequent action.

At the time, the refinery was the oldest in the midwest and the fourth largest in the United States. Its management had planned thoroughly for emergencies and had a special fire prevention committee, a plant fire marshal and a well-equipped fire brigade. Immediately after the explosion a signal on the plant fire whistle called back off-duty personnel and nearly 3,500 plant employees were available for action.

Because ground and tank fires were spreading over a large area there was no possibility of prompt extinguishment, so fire attack consisted of protecting tanks and buildings endangered by the fire. Before long, it became obvious that burning fuel flowing along the ground had to be confined, so dozens of dump trucks hastily carried loads of dirt and gravel for this purpose. Despite the diking, burning oil flowed to nearby railroad tracks and destroyed a long line of tank cars and boxcars. The City of Whiting is adjacent to the Town of East Chicago, Illinois, and mutual aid calls brought twenty-seven pieces of public fire department apparatus and 280 fire fighters from different communities. These were assigned to direct cooling streams on tanks and to protect buildings from intense radiant heat.

A "boilover" occurred during this action and thirty-one fire fighters suffered first, second and third degree burns when the oil erupted. A boilover is a violent reaction that can happen when large tanks of crude and viscous oils, containing water at the bottom, have been burning for several hours. A heat wave develops in the oil, and travels downward at a rate of fifteen to fifty inches per hour. Temperature in this oil may reach 500 to 600 degrees Fahrenheit. When this heat wave reaches the bottom water, a violent eruption can result. The beginning is usually indicated by the increased height and the brightness of the flames, followed by projection of the boiling oil. The burning oil first erupts and falls, sometimes spreading beyond the fire walls of the tank. The column of flame can be as much as 300 to 400 feet in diameter at the base and spreading wider as it reaches 1,000 feet elevation or higher.

Such tanks can also create a "slopover" in which the expanded or frothing heated oil flows or slops over the edge of the tank. The boilover is the worst situation because of the unpredictability of the size and direction of the flame column and eruption.

The Whiting plant fire was of great concern to fire officials because large quantities of flammable liquids settled into the sewer system, then flowed to a ship canal leading from Lake Michigan to the industrial area. Fortunately, a fireboat from the Chicago Fire Department was able to confine these flammable liquids. Final extinguishment of the refinery fire was not completed until eight days later.

Eruption of flame from some refinery tanks. (Elmer Budlove photo.)

Aerial view of Whiting Refinery thirty hours after fire began. Intense heat buckled railroad cars, twisted rails, and crumpled the big storage tanks. (Motorola photo.)

McKee Refinery showing spheroid tanks and other flammable liquid containers. Arrow A points to remains of the ruptured spheroid tank which was similar in design and size to Tank B.

When a flammable liquid or flammable gas is enclosed in a metal tank and then is exposed to heat, a violent reaction can be expected. Consider this story in which nineteen fire fighters were killed more than 300 feet away from the source of explosion, and the fire blast destroyed a railroad trestle and two bulldozers 1200 feet distant, the combined length of four football fields.

Tank Blast Kills Nineteen[11]

Early Sunday morning, July 29, 1956, a large spheroid tank ruptured on the property of The McKee Refinery of the Shamrock Oil and Gas Corporation about fifty-five miles north of Amarillo, Texas. Nineteen volunteer fire fighters were killed, thirty-two others were injured, including spectators, and the intense heat affected dwellings some 3,000 feet distant. The incident offered ample testimony to the extreme dangers of fighting and watching fires on refinery property.

A spheroid tank resembles a ball that has been flattened top and bottom. Tanks of this shape are used commonly in large and small refineries. There were two spheroids at the McKee Refinery plus floating roof, cone roof, and noded spheroid tanks, all well-spaced, individually diked, and of modern construction. The tanks held a variety of products including crude oil, asphalt and diesel oil. The spheroid that ruptured was forty-six feet high and contained about 500,000 gallons of a mixture of pentane and hexane. The vapors of these products are considerably heavier than air and tend to settle to the ground.

The fire was seen to originate at an open-fired heater within the diked area of an asphalt tank, about 300 feet away from the spheroid. Flames flashed back along a vapor trail to the spheroid and for the next hour this tank was involved in a ground fire, with fire also burning at its vents. The Dumas and Sun Ray Volunteer Fire Departments had responded to a call for assistance. Members of both departments, plus the refinery private fire brigade, were fighting the ground and vent fires when the blast occurred.

Apparently the fire fighters had some warning because their bodies were found 300 to 400 feet away from the wrecked spheroid. The huge flaming ball of vapors released in the rupture ignited a 20,000 barrel diesel oil tank 200 feet away and two 10,000 barrel tanks of crude oil that were 450 to 550 feet distant. A railroad trestle and two bulldozers approximately 1,200 feet away were also destroyed by the fire. Prior to the blast, observers noted that the volunteer fire fighters were directing hose streams on two tanks 150 feet away from the spheroid and that at least one hose stream was used on the tank before it ruptured.

Flame impingement from the burning vapors issuing out of the vent onto the exterior tank surface caused the metal shell of the spheroid to stretch, and eventually fail, due to the internal pressure. The metal became thin near the weld at the top; the rest of the top was sheared by the force of internal pressure. More recent experience with this type of rupture indicates that all fire fighting personnel should be removed for at least 1,000 feet from the tank and even at this distance they should be protected. Spectators should be kept at least 3,000 feet away.

School Fires [12]

One of the most practical group safety procedures ever developed is the school fire exit drill. There is no means of determining how many hundreds of thousands of children have been saved since well-planned, well-disciplined and well-supervised exit drills were first developed. There have been fatalities; incidents where smoke and flame developed rapidly to block pathways to safe exit; and panic that caused teachers and youngsters to pile up in doorways, with tragic results. Yet, through the years, school fire exit drills have worked splendidly and on thousands of occasions have brought tremendous relief to school administrators, teachers, and parents when fire emergencies occurred.

Scenes of two tragic school fires are shown in these photos. On December 1, 1958, ninety-five children and teaching Sisters were killed when fire swept through Our Lady of Angels School in Chicago, Illinois. It began as a rubbish fire in the basement, spread up an open stairway into the third floor corridor, and trapped students in their classrooms. Because of so many false alarms in the area the fire department, upon receiving the alarm, dispatched only a single pumper, a ladder truck and battalion chief rather than the normal assignment of four pumpers and two ladder trucks with appropriate complement of fire fighters and fire officers. Even so, it is doubtful that a stronger response of fire companies could have made any dif-

ference in the fatalities; the smoke, heat and other toxic products of combustion were dense and overpowering, blinding the teachers and youngsters and causing fatalities rapidly.

The fire was a traumatic shock to municipal and school authorities and citizens throughout the world. In the United States it brought careful reexamination of school buildings with the intent of modernizing their fire protection. Automatic sprinklers, automatic detection and alarm systems, and improved housekeeping were accomplished in thousands of schools within two years after this Chicago tragedy.

A few years earlier another destructive school fire had received scant attention from the public and school authorities.

In a Cheektowaga, New York, incident on March 31, 1954, fifteen children were killed in a temporary barracks-school building that included highly combustible ceilingboard.

There was no warning of the impending fire: apparently it had burned for some time before bursting into a music room in which there were thirty-one pupils, two teachers and a salesman. There was a wild scramble for escape through the windows of the one-story wood-frame building. But ten children were burned to death and five others died later in a hospital. Wood and other combustible material contributed to fire spread.

In Saint John's, New Brunswick, a small church loses its steeple and suffers complete destruction by fire. (Photos by Basil Day.)

Church Fires

There are more than 325,000 churches and synagogues in the United States and between 4,000 and 5,000 of these buildings are damaged or destroyed by fire each year. Places of worship are extremely vulnerable to fire, primarily because of their construction, and because they seldom are equipped with complete automatic fire protection.

Churches and synagogues have these characteristics in common: they are unoccupied a large part of the time; many are customarily left unlocked, and thus are open to arsonists, thieves and vandals. In the interior, these buildings have large, open spaces and combustible construction and contents; they are susceptible to troublesome heating systems, defective electric wiring, and lightning.

Incendiarism has become the leading cause of church fires, accounting for about 36.5 percent of those incidents for which the cause could be determined.

In a recent year, defective or improperly installed heating systems caused 25.5 percent of church and synagogue fires, a proportion that has not changed significantly in the past ten years.

A slightly smaller percentage (22.3 percent) of church and synagogue fires were attributed to defective electric wiring or equipment.

Lightning is a significant fire starter in these buildings, causing nearly seven percent of fires in a given year.

The principal factors that help the spread of fire in churches and synagogues are undivided open areas, combustible construction and interior finishes, concealed spaces, and combustible furnishings. Delayed alarm, delayed discovery and delay in applying water effectively in the building interior are other contributing factors.

Fire at the Pentagon [13]

Even the famous Pentagon, U. S. National Defense Headquarters, is not immune to fire. This one occurred July 2, 1959, in a ground level area of the building occupied by the Air Force Statistical Division. It required five and a half hours for control and in that time it burned a 4,000 foot square area containing three computers and between 5,000 and 7,000 rolls of magnetic tape. Property damage was estimated at nearly seven million dollars.

In the fire area, partitions were constructed of gypsum wallboard on wood studs, finished with low density combustible fiberboard. The suspended ceiling was of similar construction and featured low density combustible acoustical tile. Three IBM computing machines and associated equipment were installed on raised plywood flooring. There was a "tape vault" with the rolls of magnetic tape, plus office furniture and associated materials for the computer section. The building was not sprinklered but had small diameter standpipes and hoses with nonstandard coupling thread. A manually operated fire alarm system notified Pentagon guards and also rang an alarm at Fort Myers South Post where apparatus of the General Services Administration was stationed.

Debris in the interior of the Air Force statistical center after the Pentagon fire. (United Press International photo.)

The fire was discovered by an Air Force officer and an alarm was transmitted at 10:45 a.m. Guards and Air Force personnel tried to use extinguishers and standpipe hose but these had no effect on the fire spreading above the ceiling and in the tape vault.

Subsequent alarms brought the Arlington County Fire Department and, later, additional departments from Virginia and Maryland. One of the fire control difficulties was that fire fighters using self-contained breathing equipment, designed for thirty minute use, could only work for about ten minutes due to the length of time it required to get to and from the fire through the Pentagon's smoke-filled corridors. Consequently, a number of fire fighters tried to carry out operations without this protection and between thirty and thirty-five suffered smoke inhalation. Altogether thirty-four fire companies including seventy-one pieces of apparatus and 300 men were at this fire.

The cause was assumed to be defective or overheated wiring in the ceiling space above the computer room.

Fireground action at Arundel Park amusement hall in which eleven persons died and hundreds were injured.

A Place of Amusement

This one-story concrete block, 80- by 160-foot amusement hall in Arundel Park, Maryland, was designed for informal social gatherings. Fortunately, the designers included exit doors with panic hardware so that a large group of people could escape from the building in different directions. This, plus the fact that the exits were marked and lighted, and were of adequate width, probably saved hundreds of lives in the fire that occurred Sunday evening, January 29, 1956. But there were important deficiencies that turned the evening into tragedy.

Approximately 1,100 persons were in or just outside the hall that evening, enjoying a church-sponsored oyster roast. About 5:00 p.m., outside the building, where oysters were being cooked in a fireplace, a few men noted flames under the eaves of the roof near the fireplace chimney. Someone tried to extinguish the flames with a small garden hose but the fire spread up the roof and into the concealed undivided attic space in the area between the combustible fiberboard ceiling and the arch type wood roof. A few moments later, people inside the building saw smoke and fire coming down between the ceiling and one corner of the hall. One man grabbed a carbon dioxide extinguisher and discharged it at the flames which disappeared momentarily. The owner of the building then ordered one employee to get a step ladder and told another employee to call the fire department. The first alarm was received at fire headquarters at 5:08 p.m., approximately eight minutes after first discovery.

The employee who opened the trap door and looked into the attic space saw that fire apparently involved the whole area so he quickly shut the trap door and the building owner then ordered everybody out of the hall.

As people started toward the exits a "swoosh" sound was heard and fire and smoke boiled through the ceilings in several places. Immediately there was a rush for all available exits and the windows. The main hall had three 3-foot exits leading directly outside; another 3-foot exit leading to the main entrance (a six foot double door); and two six foot exits, one leading to the milk bar and the other to the cocktail lounge. These were all wooden doors, equipped with panic hardware. There was also a four-foot door leading directly outside from an attached kitchen and two overhead garage-type doors, one at each end of the hall.

Unfortunately, the floor of the hall was covered by many loose chairs and tables, used for single parties and other game functions. These helped to block passage of the people trying to escape. Although fire companies responded immediately to the alarm, hundreds of people who had escaped from the building were milling around the area, dazed by injury, panic or confusion. This made fire fighting and rescue efforts very difficult. Because of the tremendous heat fire fighters could not enter the building for more than two hours. Ten women died in the fire, a man died of burns a week later and hundreds of other persons received burns or injuries.

Factors which contributed to the loss of life were: failure to evacuate the building immediately when fire was discovered; failure to call the fire department promptly; congested conditions in the hall, aggravated by loose chairs and tables, and lack of emergency lighting.

The H-Bomb

One of the great moments in scientific history received little coverage from the general media at the time. In 1939 the famous scientist, Niels Bohr of Copenhagen, Denmark, was in Princeton, New Jersey, at a conference. He announced that two other scientists, Lise Meittner and her nephew Otto R. Frisch had interpreted results of research by two German scientists, Otto Hahn and Fritz Strassman as being the fusion of uranium atoms bombarded with neutrons that resulted in a tremendous release of energy in the form of light and heat. This somewhat informal announcement immediately sent scientists throughout the United States and Canada into a frenzy of research and the atomic age was beginning.

In the Fall of 1939, because of the significant developments of war in Europe, the scientist Albert Einstein was persuaded to send a letter to President Franklin D. Roosevelt urging that the U. S. government should support research in the application of atomic energy. Step by step, this led to temporary military control of atomic energy, research and development. The famous Manhattan Project was established in 1942;

simultaneously in universities and research centers throughout the country work was proceeding at a swift pace to unveil or uncover the tremendous capabilities of atomic energy. On July 16, 1945, the first plutonium atomic bomb was exploded at the testing site in Alamogordo, New Mexico. Shortly after this, the two atomic bombs of 20,000-kiloton capacity were dropped on Hiroshima and Nagasaki, Japan. These were not of the plutonium type; they were made of the untested U-235 element, but proved successful.

Today, hydrogen and other atomic bombs are 2,000 to 3,000 times as powerful as those first two weapons used in World War II.

This was one of the countless tests by the U.S. from the 1950s to the mid-1970s. This photo was taken at an altitude of about 12,000 feet, at fifty miles from the detonation site. Two minutes after Zero Hour the cloud had risen to 40,000 feet. Ten minutes later, as it neared its maximum, the cloud stem had pushed upward twenty-five miles into the stratosphere. The mushroom portion went up ten miles, spread for 100 miles. (*U.S. Air Force Photo.*)

211

Fire Fighter Donald J. Stemo of the Village of Whitefish Bay, Wisconsin won an award from the Eastman Kodak Company for this 1975 newsphoto of a plane crashing and igniting. Note pilot ejecting at left. (Eastman Kodak Company contest photo.)

Aircraft and Airports

Today's aircraft, used for passenger and freight transportation, are the safest ever developed. The aviation industry offers convincing statistics comparing the favorable accident record of air travel with other forms of public transportation, and improvements continue to be made. This is commendable, because air travel is and will continue to be one of our most popular forms of transportation.

It is a fact that aircraft have many inherent fire hazards; fires in years gone by have underscored the need to provide them with maximum fire protection. In the air, passengers must depend upon the intelligence and skills of the pilot and crew, and the designed fire protection arrangements within the plane; on the ground, passengers rely upon the experience and foresight of the terminal's management and airport fire brigades to provide maximum protection against accidental fires.

Aircraft have these obvious fire potentials: A large quantity of flammable or combustible liquid used as fuel; a significant amount of oxygen that can accelerate a fire condition; combustible cabin materials, some belonging to passengers, and certain extras necessary for passenger service; a fuselage covering, or "skin," that will deteriorate quickly in fire conditions; tires and hydraulic fluids that may contribute to the fire problem.

Possible sources of ignition on a plane include: electrical circuits; static and friction sparks; smoking materials; lightning; human accidents; and the derangement of the components as a consequence of a hard landing or crash impact.

During the past ninety years the aviation industry had become very aware of the devastating effects of fire in flight or on the ground. The modern commercial aircraft has a well-trained highly-skilled crew; in-flight fire protection is provided for the power plants and some cargo spaces, plus portable fire extinguishers in the cabins. Protection is designed for fuel, oxygen and electrical systems to minimize the fire potentials each present. A standard set of safety procedures is used to inform passengers at the start of each flight in the event of an emergency. Mechanically operated evacuation slides are provided to assist in rapid escape when a plane is on the ground.

There are fire hazards at airports also: aircraft maintenance and servicing operations; fuel servicing and fueling operations; possible collision impact with another airplane, or a ground vehicle; careless smoking by workmen or others in the vicinity; fuel spills; combustible materials — paper, wood, plastics, upholstery — that may be ignited by normal "accidents." Hangars are susceptible to large fires; passenger and freight terminal buildings have also been involved.

Under Federal regulations, airports are required to provide a certain amount of fire apparatus and fire fighting personnel according to the frequency of flights and the type of aircraft using the terminal. Fire control crews, chosen according to their abilities and qualifications, use special crash and rescue vehicles and other equipment developed for the aircraft fire.

The Vulnerable Interior[1]

The vulnerability of aircraft interiors to fire damage, even without involvement of the propulsion fuels aboard, is illustrated by this fire which occurred April 19, 1974. The aircraft, an L-1011, was left unattended after a post-flight inspection following an 8:00 p.m. arrival at Boston's Logan Airport. About midnight, an employee smelled smoke, boarded the aircraft and discovered smoke coming from the aft end of the fuselage. He alerted the airport fire department. Inaccessibility of the smoke-clogged interior and a burn-through at the top of the fuselage near the No. 3 engine of the high aircraft prevented effective fire fighting, except for protecting the fuel containing wing structures. The passenger cabin area was completely destroyed and heavy damage was caused to the cockpit area of the $22,000,000 aircraft. The Boston Fire Department made good application of foam and an elevated stream from an articulating water tower. (*Boston Globe* photo.)

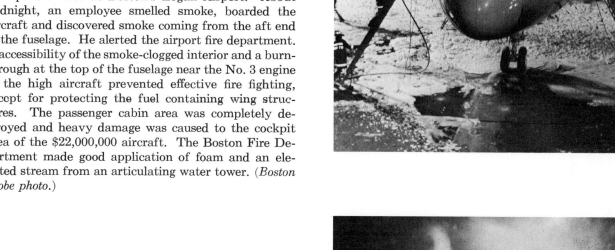

A Convincing Need for Crash Fire Apparatus[2]

Twelve persons were killed in a DC-4 accident at the Chicago Municipal Airport on March 10, 1948. At least seven died as a result of the ensuing fire; the others presumably because of impact injuries. The aircraft struck the ground with power off after having stalled at approximately 600- to 800-foot altitude. About 1,800 gallons of fuel fed the fire almost instantaneously after impact. At the time, there was no fire equipment stationed at the airport; the nearest Chicago Fire Department engine company had to travel 1½ miles to reach the scene; others had to travel greater distances. Within a couple of months, the department ordered its first specialized aircraft rescue and fire fighting vehicles for the airport — the world's busiest. (*Associated Press* photo.)

Outboard Fuel Tanks and a Bogged-down[3] Crash Truck

Thirty-four occupants of this Convair 240 escaped unaided after the accident at the Tulsa Municipal Airport, February 27, 1951. Fire started on, or shortly after, first impact. Two reasons why escape was possible were that impact forces were not fatal and wind direction was favorable, but the main reason was that the fuel tanks were located outboard of the engine nacelles and flames did not enter the fuselage with the speed experienced where the fuel is in the wing center sections. No fire fighting was accomplished because the airport-based crash truck bogged down in soft ground after getting about 100 feet off the runway's paved surfaces. The aircraft came to rest 900 feet further away. (*Photo by Chester Sharp.*)

A Horizontal Chimney[4]

"Crumpling of the main landing gears on the ditch banks resulted in rupture of the inboard fuel tanks, since the gear was structurally supported by the front and center wing spars which also supported the inboard (main) integral fuel tanks. The presence of such a ditch directly at the end of the runway was most unfortunate." (It had been constructed to depress a steel airport boundary fence because of the proximity of the ILS (instrument landing station) localizer. This tells the basic story of the DC-4 accident at Philadelphia International Airport on January 14, 1951. Seven lives of the twenty-eight occupants were lost. The fire spread into the fuselage quickly and traveled down its length, because there was no fire barrier at the wing root (where the wing passes through the fuselage). The opening to the main cabin door for exit purposes created a "chimney" effect in the cabin. (*Wide World photo.*)

It's Nice to have Fire Fighters Nearby[5]

Collisions between aircraft on the movement areas of airports always present a fire potential. In this November 10, 1958 incident a Constellation veered off Runway 31R at New York International Airport during an attempted take-off. After crossing three taxiways, it crashed into a Viscount aircraft parked at a temporary terminal building. The Constellation was in flames during its erratic course. The port wing had made contact with the ground (because of a ground depression) between the taxiways and this resulted in the wing tearing off, so that when it struck the other aircraft a major fire occurred involving fuel in starboard tanks of the Viscount. The airport fire department was able to protect the escape route of the five crewmen aboard the Constellation, reduce the fire intensity by foam coverage of the hulls and ground spills, prevent involvement of the fuel in the port tanks of the Viscount 600 to 800 gallons of JP-4), and to protect the terminal. (*Wide World photo.*)

The Problem of Oxygen Under Pressure[6]

Oxygen-fed fires can cause major aircraft damage. This fire was one of the first which caused extensive damage to a DC-4 at LaGuardia Airport in New York on September 28, 1948. It started from an internal explosion of an oxygen line during preflight servicing. Other serious ones occurred on January 24, 1956, involving a Stratocruiser at London, on July 14, 1959, at Wold-Chamberlain Airport, Minnesota, also involving a Stratocruiser, on December 1, 1962, involving a 707 at Bombay, India, on November 16, 1963, involving a 720 at Miami, on September 7, 1968, at Rio de Janeiro destroying a 707 and two fires involving 737s, one on January 25, 1969, at Boeing Field and the other on December 31, 1970, at Washington National Airport. (*Acme photo.*)

214

Tire blow-outs and a narrow aisle[7]

A DC-8 carrying 115 passengers and a crew of seven collided with a parked truck near to taxiway construction area at Denver's Stapleton Field July 11, 1961. The collision followed blowout of some of the eight tires on the main landing gear. Fire erupted immediately and while most occupants were able to escape, sixteen died of asphyxiation in the smoke-filled cabin. The narrow aisle width (15½ inches) was cited as one reason why cabin attendants could not accelerate the escape of the victims. (*Wide World photo.*)

Fire-in-Flight[8]

Fire-in-flight occurred on this 707–320B with 153 persons aboard enroute from San Francisco to Honolulu June 28, 1965. Built-in design safeguards allowed Engine No. 4 (which had an internal malfunction) to break away and fall free. Despite continuing fire, suppressed in part by closure of a fuel valve to the affected area, the pilot was able to make a safe landing at Travis Air Force Base, twenty-three minutes after take-off. The fact that the pilot could shut off the flow of fuel by closing the valve near his cockpit seat was another planned safety feature that limited fire spread. Boeing also designed single pods, rather than dual pods, for this aircraft so that a single engine failure would not have a serious operational effect on an airborne aircraft.

Some Escaped, Despite Fuel Spray[9]

This accident at Salt Lake City is one of the most significant in the history of commercial aviation as far as crash-fire factors are concerned. It occurred November 11, 1965 and involved a Boeing 727. After a faulty landing (hard and short of Runway 34L) the right land gear, wheels and main strut were pulled into the fuselage between the wing trailing edge and the intake of the right engine. When the gear punctured the fuselage, the main cabin floor and its supporting members were buckled, but worse than that, the fuel lines supplying No. 2 and 3 engines and the alternating current power buses from the engine driven generators in the "underbelly", were shattered. This allowed fuel (Jet-A-kerosene) pressurized to 35 psi to flow into the cargo pit and the vapors ignited from the severed electrical buses. The consequent fire burst into the cabin and the result was the death of forty-three occupants, but in about thirty-seven seconds after the plane landed, passengers were escaping as the airport crash crew was responding. (*Salt Lake Tribune photo.*)

215

On August 8, 1972, in Los Angeles, California, this tank containing flammable liquid rocketed hundreds of feet through the air above heads of fire fighters. It landed on the roof of building at lower right between two groups of fire fighters. (Jack Wyman photo.)

"... I believe that this nation should commit itself to achieving the goal, before this decade is out, of landing a man on the moon and returning him safely to earth. No single space project in this period will be more impressive to mankind or more important for the long-range exploration of space. And none will be so difficult or expensive to accomplish."[1]

President John F. Kennedy in a 1961 speech to the U.S. Congress.

1961-1976

Chapter VIII

With the beginning of the "Golden Sixties" the United States was starting a new era of technological achievement unequalled in any previous period of history. It was committed to a space program that, as promised in 1960 by President John F. Kennedy, would bring landings on the moon and the launching of hundreds of space satellites and space probes. The research developed as part of this program had some influence on fire protection equipment and techniques, particularly with respect to communication, solid state technology, data gathering and processing, and the use of synthetic materials. At the height of this, a needless, tragic flareup of oxygen in a space capsule took the lives of three astronauts and created a humbling lesson of the dangers of fire to scientists, technologists, and the general public.

Terrible loss of life incidents in modern high-rise buildings in North and South America, Europe and Asia, captured the attention of the world. The fact that occupants of these huge, vertical complexes could be trapped, injured and killed by fire startled the public and responsible authorities, particularly when these incidents were flashed instantaneously throughout the world by the immediacy of on-the-scene television. Fire tragedies in Seoul, Korea; Sao Paulo, Brazil; New Orleans, Louisiana; New York City, New York; London, England; became a great challenge to the fire protection profession and industry and led to strengthening of laws and regulations pertaining to buildings of public occupancy. In addition, they brought a new use of helicopters as a means of rescuing people trapped in these buildings, and bringing fire fighters and equipment to positions for tactical control of the fire.

"BLEVE" became a new acronym in the terminology of fire control. A "BLEVE" occurs when flame impinges on the vapor space of a tank containing flammable liquid or gas, and the result can be a tremendous explosion. A few of these incidents are described in the following pages.

Severe incidents of fire and explosion occurred in each of the principal means of transportation: in the air, on land, and in the water, and they involved a variety of dangerous gases, liquids, and solids.

The helplessness of elderly patients in nursing homes, hospitals and homes for the aged was dramatized during the 60s by a tragic series of loss of life fires. As a result, legislation requiring automatic sprinkler systems and other improved fire protection for these occupancies was passed on Federal and state levels.

The grim history of mining continued its record of tragedy in the third quarter of the twentieth century with miners and other workers trapped, burned, and killed in fires, explosions and mine collapses. (*See page 154.*)

The fire experience in the United States continued to exceed that of any other country, with an average annual loss of life of between 11,500 and 12,500 persons, and a monetary property loss estimated in excess of four billion dollars. But to counter this deplorable experience, the science of fire protection, prevention and control improved with new extinguishing agents, new methods of detection, new apparatus and equipment and a growing mass of information and technology, creating the most intelligent, the most competent actions against fire we have ever achieved.

Bel-Air Conflagration[2]

One of the most costly conflagrations of the twentieth century occurred in the brush covered hills of Los Angeles, California, November 6, 1961. It destroyed or damaged more than 450 expensive homes in the Bel-Air and Brentwood residential section of the city and caused a financial loss estimated in the vicinity of twenty-five million dollars. It was the worst conflagration to occur in North America since a brush fire in Berkeley, California, September 17, 1923, caused destruction of 640 buildings.

As in that Berkeley conflagration, combustible dwellings with wood shingle and shake roofs were a major factor in development of this Bel-Air fire from a difficult brush fire into a sweeping conflagration that was beyond reasonably quick control.

In most of California, but particularly in Southern California, fire departments receive plenty of experience in fighting brush and forest fires. These become prevalent and dangerous in the Fall of each year. When the combination of strong winds, low humidity and dying vegetation creates tinderbox conditions. One of the most dreaded situations is when the high velocity "Santa Ana" winds from the northeast sweep down through the mountains in the vicinity of Los Angeles.

This Bel-Air Conflagration began at approximately 8:10 a.m. near a canyon road on the north side of the Santa Monica mountain range. Because strong wind conditions had prevailed since the preceding Saturday, the Los Angeles City Fire Department had already begun to move fire companies

This fire in Mint Canyon, 40 miles northwest of Los Angeles, shows how intense flame races through brush in the steep terrain. It is very difficult to get tankers and other fire apparatus into effective positions for such fires. (Wide World photo.)

to strategic positions to supplement existing protection in the mountain area. However, with a northerly surface wind gusting at times in excess of 35 mph velocity, fire developed rapidly. Humidity had been below ten percent since Sunday afternoon.

Fire companies responding to the first telephone alarm managed to control the flanks of the fire, protecting the dwellings first threatened. But the broad fire front spread up canyon walls and leaped across the ridge road, Mulholland Drive. In its path were the homes of some 4,000 people and the immediate task was to warn these people and evacuate them from the threatened area.

Many of these expensive dwellings, built on steep hills, were practically enveloped by brush and trees. Chaparral brush and other low-growing species in the area, when heated, emit oily vapors that ignite almost explosively. In a canyon or on a hillside the flareup of such ignition is almost instantaneous. Even the efforts of the two major professional fire departments — Los Angeles City and Los Angeles County — were not enough to stop such intense, rapid spread.

To make things worse, burning embers from the brush easily set fires down wind (south) in other brush and on combustible roofs. When the roofing became involved, large flaming brands were carried aloft by thermal currents and along the path of upper winds. At about 2,000 to 3,000 feet elevation, strong cross winds blowing to the southwest carried these brands to places more than a mile to the side of the main fire. Some of these brands caused single buildings to burn to the ground while other buildings and vegetation nearby were hardly touched by the fire. Many minor new fires were extinguished by fire fighters or homeowners, but other dwellings, in similar conditions, burned beyond control.

This conflagration brought full response from the Los Angeles City and County Fire Departments. More than 3,100 fire fighters and about 240 vehicles were in action, while twelve airplanes and five helicopters operated to reconnoiter the fire and drop water with fire retardants. Apparatus from twenty-four other municipal fire departments covered stations in Los Angeles while the conflagration was underway. Altogether the brush fire extended over 5,000 acres and more than 450 homes and 180 other buildings were destroyed. Of the dwellings more than 300 had roofs made of wood shingles or wood shakes. Brush fires continue to be a major threat to California dwellings.

Remains of nursing home in Fitchville, Ohio where sixty-three elderly patients died in a fire on November 23, 1963.

Nursing Homes [3]

In our present society it should be obvious that residents of nursing homes and custodial care occupancies may be assumed to be physically and mentally helpless in any sudden emergency. It is quite likely that these persons are aged, or physically or mentally infirm, so that if a fire or any other life-threatening emergency occurs, they need a great deal of assistance from staff personnel.

In the twentieth century, the record of fires in these institutions has indicated the need for maximum fire protection — automatic alarms, a sufficient number of well-designed and well-lighted exits, a minimum of combustible interior material, complete protection by automatic sprinklers, well-planned layout of corridors, interior doors, service area and hazardous areas plus adequate alarm detection and extinguishment system designed to fulfill specific needs.

Staff personnel need complete training in how to evacuate patients and other persons from the building, how to send an alarm, how to communicate a fire emergency to administrative personnel and to other responsible persons, and how to move helpless patients, whether or not means of easy transportation is available. As in school fire exit drills, and in similar drills for dwellings, hospitals, or other occupancies, it is essential to have a basic building evacuation procedure with an assembly point outside where all staff personnel and building occupants can be identified. The spread of fire, smoke and other by-products can incapacitate aged, infirm or retarded persons very quickly because in their normal routine they are dependent upon the staff personnel who care for them.

A survey in the mid 1970s indicated that multiple-death fires in homes for the aged caused 338 fatalities within twelve years. There were 6,100 fires in nursing home occupancies in the most recent year. Among the most serious incidents were a Marietta, Ohio, convalescent home fire on January 9, 1970 that killed thirty-one; a fire in Philadelphia that took the lives of eleven persons in September, 1973; a fire in a home for the aged in Benchel, Kentucky killed ten; a nursing home fire in Honesdale, Pennsylvania, that killed fifteen on October 19, 1971; a fire in Kearney, Nebraska, killed four nursing home residents, injured three other residents and a police officer; and a fire in a "skilled nursing home" in Wayne, Pennsylvania, took the lives of ten aged victims.

220

Scene at Houston fire incident just before the vinyl chloride blast occurred. Note fire fighter at top of aerial ladder. Newsmen and others without protective clothing were severely burned when the tank ruptured.

Vinyl Chloride Blast[4]

On October 19, 1971, officers and men of the Houston, Texas Fire Department were stunned and injured by a violent blast from a fire-involved railroad car containing vinyl chloride. The fire companies had responded to a train wreck and were attempting to direct cooling streams on the critical part of the fire. When the explosion occurred four chief officers in command of the scene and twenty-one company officers and fire fighters were knocked out of action. One training officer was killed, fourteen other men were injured and a fire department pumper was destroyed. Five newsmen were taken to a hospital with burns or other injuries.

This blast was similar to the BLEVE described on page 226, in that one of the railroad tank cars ruptured violently and created a fireball which witnesses stated was more than 1,000 feet in diameter. The surrounding area was showered with parts of tank cars, rocks, cross ties and flying debris caught up in the massive release of energy. The ruptured tank was later identified as having a cargo of 48,000 gallons of vinyl chloride.

Most of the fire fighters had just enough time to turn and run when they noted an accelerated build-up of the fire just seconds before the rupture. As a result, they received most of their burns on their backs and hands. However, one fire fighter who climbed an aerial ladder to operate a ladder pipe was caught on top of the ladder with the full force of the fireball and blast. Although he practically dived down the ladder, his protective clothing was in shreds, his hip boots had melted rubber, his helmet was badly charred and he was severely burned. A nearby member of the fire department training division was killed in the blast.

Vinyl chloride at ordinary temperatures is a colorless sweet smelling flammable gas. It becomes a liquid below 7 degrees Fahrenheit. It is usually shipped as a liquid in cylinders, tank cars or tank barges. It is a skin irritant and toxic, since it acts as an anesthetic and can be fatal in high concentrations. When fires involve vinyl chloride, highly toxic hydrogen chloride and carbon monoxide can result. This devastating blast in Houston was similar to another incident that occurred in Glendora, Mississippi, September 11, 1969. There a tank car of vinyl chloride was exposed to fire for about ten hours before the tank rupture occurred.

The violence of tank car rupture when certain commodities are involved by fire was demonstrated by an incident in Laurel, Mississippi, where a thirty-seven foot section of a tank car was hurled through the air for more than 1,600 feet. Fire fighters are aware of these dangers but cannot always anticipate that the worst situation is about to happen.

221

Boston

Portland, Oregon

Riots and Civil Disorders [5]

The decade of the 1960s was also noted for a large number of riots and civil disorders in which thousands of fires were started deliberately and fire fighters and police officers became victims of shootings and other forms of assault. The first major disorder occurred in the Watts section of Los Angeles, California, in 1965 and during the next three or four years many other major U. S. cities experienced this type of violence.

The Watts riot began in the southern part of Los Angeles in the early evening of August 11, 1965, when a local citizen was arrested for driving under the influence of liquor. Within two days this rioting burst into a tremendous outbreak of vandalism, burglary and looting. Automobiles were overturned, store windows were broken, liquor and guns were taken wherever they could be found, and sporting goods, television sets, food and other articles were stolen. A number of fires were started by the looters and, when Los Angeles fire fighters responded, bricks, stones, sticks and even Molotov cocktails were thrown at the apparatus. Four fire

fighters were shot (only two with serious wounds) and 183 sustained injury. A total of 104 fire trucks were damaged and forty-five had their windshields broken.

In one twelve-hour period the Los Angeles Emergency Control Center received over 3,000 alarms and it was estimated that at least 2,000 fires occurred during the three days of rioting. On Friday evening August 13, it was reported that at least 200 fires occurred simultaneously within the city.

Days later, when the rioting had subsided, it was estimated that about forty-seven square miles had been affected. Within city limits 261 buildings housing 318 separate occupancies were partially or totally destroyed. In the surrounding area of Los Angeles County more than 600 buildings were damaged by burning and looting, and over 200 of these were totally destroyed by fire. The fire loss was estimated at more than two million dollars. The number of fatalities was estimated at thirty-four and the number injured at 1,052. [6]

222

From July 12 to 17, 1967, Newark, New Jersey, had a major disturbance that included widespread rioting, sniping and arson. Twenty-three persons died including a police officer and a fire captain. Approximately 2,000 buildings were damaged with more than 100 involved by fire.

A week later Detroit, Michigan, went through a similar experience in which 538 business establishments were totally destroyed and 549 others suffered serious damage in the eight days between July 23 and July 31. One fire fighter was killed by a bullet, and at least thirty-eight persons died and more than 1,500 persons were injured. Property loss in Detroit was estimated at more than half a billion dollars.

In the middle of these tragedies the Congress of the United States passed special legislation making it a Federal offense to interfere with fire fighters and police officers in performance of their duty in these disturbances.

The City of Philadelphia had a major civil disturbance in August, 1964, in which property damage totaled over three million dollars but fire was not a major problem.

Boston, Massachusetts, had a series of disturbances during June of 1967 when false alarms became so numerous that no fire companies were immediately available for response to actual fires.

These tragic disturbances finally began to subside in 1968 but they left a serious aftermath. Ever since, fire fighters in many localities have suffered provocative attacks when they have responded to alarms in certain areas. Because of assault by bottles, stones and other missiles, it has been necessary in some cities to redesign apparatus and protective equipment for the protection of fire fighters. Some fire departments adopted a new

Milwaukee

"crash" type of helmet; others developed special steel enclosures for ladder trucks and other apparatus in which personnel were exposed to thrown objects and other dangers. On March 1, 1968, the National Advisory Commission on Civil Disorders issued its report on twenty-three cities studied by the Commission. Most of these cities reported arson and fires accompanying the disorders of the '60s. The report identified the major fire department problems in these emergencies and included recommendations for improving fire department safety and defense.

In Sylva, North Carolina on April 20, 1958 fire started in the unsprinklered linen room of a two story brick, wood joisted hospital. The hospital staff used portable fire extinguishers without success, then removed all patients from the building while fire fighters gained control and extinguished the blaze.

(Sylva Herald photo.)

Fires In Hospitals [7]

A five-year survey at the end of the 1960s indicated that an average of 4,180 fires in hospitals occurred each year in the United States. A few years later another study indicated that hospital fires had leaped to 15,600 in 1974 causing damage in excess of twenty million dollars. Regardless of the discrepancy in these surveys, it is obvious that hospitals are easy victims for accidental fires. Since many of the patients are bed-ridden or otherwise confined there is a high potential for injury, burns and loss of life in any well-developed fire.

Subsequent to World War II many hospitals had improved their fire protection, partly because of the terrible fire incidents that occurred in hospitals and institutions before 1940 and because improvement of medical equipment and hospital operating procedures, together with the increased use of flammable gas, practically forced improvement of building design and fire protection. Many hospitals installed automatic detection, alarm and extinguishing systems and otherwise improved building layout, interior materials, training of personnel and other common safety practices to minimize the danger of fire. Nevertheless, during the 1960s there were several bad examples of hospital fires.

On December 8, 1961, fire started in the principal hospital in Hartford, Connecticut, a building considered to be safe from large loss of life fires because it was of modern fire-resistive construction of that time. It was not anticipated that heat and smoke produced by the burning of interior finish could be quite as effective in killing people as fire products from other wooden materials.

The Hartford Hospital was a thirteen-story fire-resistive building that had been completed in 1948. Partitions were made of utility tile and plaster with linoleum wainscotting and a plastic covered fabric finish above. In all thirteen stories the corridor ceiling was made of combustible fiberboard acoustical tile, mounted on gypsum lath by an adhesive. The floors were covered with linoleum but there were single-swing metal clad doors in metal frames at each end of the center section corridor, and these were designed to hold back the spread of fire and smoke.

The hospital had a metal rubbish chute approximately twenty-two inches in diameter that extended from the sub-basement to the roof with a three-inch vent pipe at the top and one sprinkler on each floor. This chute opened directly onto the corridor by an aluminum door (not a standard fire door).

Fire started in the rubbish chute during the lunch period of one of the maintenance personnel. When he discovered it on his return he tried vainly for several minutes to extinguish it but burning continued. The mass of rubbish in the chute apparently prevented operation of a single sprinkler, but finally black smoke began to seep out on upper floors and a nurse on the twelfth floor pulled a fire alarm box. At that time there were 793 patients in the hospital, 108 of them on the ninth floor plus some one hundred employees and visitors on the same floor. For some reason the chute door on this floor flew open and flames poured into the hospital corridor at 2:40 p.m. One minute later the sprinkler operated transmitting an immediate alarm to fire department headquarters.

The blast of fire from the chute opening ignited the combustible ceiling tile in the corridor and flames poured down the hallway "as if from a flame thrower."

Hospital personnel responding to the emergency closed doors to stairways retarding the spread of

Eighty-nine patients were removed from this fire-resistive hospital in Des Moines, Iowa on January 29, 1965. A workman's cutting torch had ignited combustibles in the old two-story wing of the building and the fire spread up through nonfirestopped walls and ceilings. The new section was protected by fire doors but heavy smoke forced evacuation of the building. (Wide World photo.)

smoke and preventing penetration of the fire to other floors. Arriving fire fighters ventilated upper floors and, from the tops of ladders, convinced patients and other people of the ninth floor to close their doors and to use wet bed clothing at cracks. Those who followed this advice escaped unharmed. Where doors to patients' rooms did not stay closed the occupants died. Altogether seven patients, four employees and five visitors — a total of sixteen — died on the ninth floor of this building.

After the fire, samples of the ceiling tile were tested and were found to have flame spread rating of 180 when applied with adhesive to gypsum board. The linoleum wainscotting had a flame spread rating of 300. Both of these numbers indicate potential of rapid flame movement over the combustible material.

On April 17, 1960 a major hospital in Memphis, Tennessee was exposed to heat of a fire that was burning the grandstand of a baseball stadium. The wooden window frames of the hospital maternity ward ignited by radiant heat at a distance of forty-five feet across an avenue. The windows broke and nurses removed the infants from the ward but did not disturb the mothers who were not endangered.

A different part of the hospital was more affected by the heat which later was pushed by strong wind gusts. Ordinary plate glass windows on the first floor broke, allowing fire to enter and ignite furniture and draperies in the hospital first-floor lobby. Eventually about 250 patients had to be moved to different wings of the hospital while the Memphis Fire Department extinguished the fire.

Another fire of the 60s was the one which involved the New Mexico State Hospital in Las Vegas, September 22, 1967. Six patients died when fire raced through the combustible interior and contents.

Flammables in hospital laboratories, oxygen, and anesthetics have caused injury and loss of life. Sparks from static and electrical appliances have been sources of ignition for flammable gas incidents. Smoking materials have also contributed to some of the tragic incidents.

The Terrible Blast of a BLEVE![8]

In fireground operations there are three sudden changes every fire fighter dreads: the unexpected interruption of waterflow from a nozzle; the first sounds and signs of impending building collapse; and the sudden, unexpected violence of an explosion.

Loss of water can be dangerous, especially if one or more fire fighters are in a position where they may be trapped by flame, smoke and other products of combustion. A building collapse can also be dangerous but usually the condition of the building, or previous information of the structure gives warning of danger. An explosion may occur with no previous warning and the direction and strength of its force may determine whether the fire fighters live or die. A BLEVE falls into the category of dreaded explosions!

BLEVE is a new term in fire control, developed after several major fireground incidents. It is an acronym standing for Boiling-Liquid-Expanding-Vapor-Explosion. It is a term that should be understood by most fire fighters and hopefully, someday, it may serve as a warning to enthusiastic fire-watchers who are inclined to move close to fire situations.

The incident shown in photos at left occurred the night of February 9, 1972, in Tewksbury, Massachusetts, when a propane transport truck ruptured. Two men were killed and twenty-one were injured. Damage to a nearby propane and liquefied natural gas operating facility was about $150,000; about $70,000 in fire apparatus and equipment.

The explosion occurred in typical conditions of a BLEVE: liquid propane was burning at a discharge line break and the flame was impinging on the cargo tank of the truck. The Tewksbury Fire Department had responded to a telephone alarm and fire fighters were operating a deluge gun and a handline, attempting to keep the tanks cool. After the fire had been burning for about twenty minutes the cargo tank ruptured violently, and came apart in two sections. The rear portion

226

flattened out and its contents spewed laterally in nearly a 360 degree arc. As a result, fire fighters operating a deluge gun about 150 feet away and a handline 120 feet away were knocked around and burned.

The front portion of the tank, in cylindrical form, still attached to the tractor, was propelled forward, jackknifing the tractor-trailer combination. The entire combination moved along the ground for about thirty feet where it smashed into a vaporizer. The tank portion then became airborne and sailed through the air. After shearing off three eight-inch diameter trees several feet above the ground, it came to rest about three hundred feet from its original position on the truck.

Later, it was estimated that of 6,500 gallons of propane in the tank, 3,600 flash-vaporized, mixed with air and was ignited — creating a large ball of fire. The remaining cold propane was atomized and was flung in all directions, in burning and unburned form.

Here's why these explosions develop so much power:

Propane in a container at any temperature above its normal boiling point of about minus 45 degrees Fahrenheit, contains sensible heat which is immediately available for vaporization of the liquid if the container pressure is reduced to atmospheric pressure. At 70 degrees Fahrenheit, this heat is sufficient to vaporize almost instantly about 36 percent of the liquid in the container. At 130 degrees Fahrenheit (the approximate temperature of the liquid corresponding to 250 psi vapor pressure — the initial discharge pressure of a relief valve on a propane storage tank or transport cargo tank), this heat is sufficient to almost instantly vaporize about fifty-five percent of the liquid in the container.

The remaining unvaporized propane is refrigerated by the "self-extraction" of heat when the pressure is reduced to atmospheric and cooled to near its normal boiling point.

Liquid vaporization is accompanied by a liquid-to-vapor expansion of more than 270 to 1 — that is, one gallon of liquid will expand to about 270 gallons (36 cubic feet) of propane vapor. It is this expansion process which provides the energy for crack propagation in the container structure, propulsion of pieces of the container, rapid mixing of the vapor and air resulting in the characteristic fireball upon ignition by the fire which caused the BLEVE and atomization of the remaining cold propane liquid. Many of the atomized droplets burn as they fly through the air. However, it is not uncommon for the cold liquid to be propelled from the fire zone too fast for ignition to occur and fall to earth still in liquid form. In this case, dissolved spots in asphalt paving were noted up to one half mile from the BLEVE site. In other BLEVEs, firefighters have been cooled by cold liquid passage in their vicinity.

Reduction of container internal pressure to atmospheric level results from structural failure of the container. This failure is most often due to weakening of the container metal from flame contact as in the present instance. (*This is why hose streams should be directed on this point of flame contact, if possible — Ed.*) However, it will happen if the container is punctured or fails for any other reason.

In twenty years prior to 1970 there were at least eighteen incidents of BLEVEs resulting from LP-Gas tanks being exposed to fire. Those eighteen fires led to the death of two fire fighters and twenty civilians with injuries to 318 fire fighters and civilians. In contrast, in the years between 1970 and 1975 there were twelve BLEVE incidents resulting in the deaths of eighteen fire fighters and six civilians and injuries to more than 300 persons.

Here are three of these tragedies:

On July 5, 1973, a propane tank car on a railroad siding in Kingman, Arizona, caught fire and exploded. Twelve fire fighters and a civilian were killed from burns and another was in critical condition. Ninety-five persons were injured, most of them spectators clustered along the highway about 1,000 feet from the explosion, who ignored police orders to move back.

On January 11, 1974, in West St. Paul, Minnesota, fire erupted when an 11,000 gallon LP-Gas stationary storage tank was being refilled near an apartment building. Heavy fire impingement on the tank shell above the liquid level resulted in a BLEVE which killed three fire fighters and one civilian. Two apartment houses were destroyed in the complex served by the tank.

In Oneonta, New York, on February 12, 1974, fire developed after a railroad freight train derailed, involving twenty-seven cars, seven of which carried LP-Gas tanks. BLEVEs occurred in four of these tank cars and portions of one tank flew through the air more than 1,300 feet. More than fifty people were injured, most of them fire fighters.

Fires In Tall Buildings[9]

In the latter part of the 1960s and the first half of the 1970s fires in "high-rise," or tall buildings became a very serious problem. They occurred with alarming frequency, sometimes trapping people, with consequent loss of life. Despite the fact that most of these fires happened in buildings of the latest design, they quickly focused attention on these inadequacies: They developed on floors that were far above the immediate capabilities of local fire apparatus and equipment. They magnified the importance of early fire detection and alarm signaling to the local fire department; they underscored the need for adequate exits on each floor, designed to give all persons a reasonable travel distance to a smokeproof tower, or other safe location. They also called attention to the fact that air conditioning systems and other air movement equipment can spread fire and toxic products of combustion. They raised a number of questions for architects, fire protection engineers, the building industry, and municipal authorities, principally concerning regulation of building design and the use of combustible furnishings and other material inside these structures, some of which are occupied by thousands of people.

A brief summary of high-rise fires which occurred in recent years helps to identify the typical situations that occurred in some of these buildings.

In Montgomery, Alabama, on February 7, 1967, twenty-five persons died in a penthouse restaurant on top of a ten-story building, the largest loss of life restaurant fire in the U. S. since the Cocoanut Grove tragedy in Boston in 1942. (*See page 180.*) This building was relatively small compared to more recent high-rise buildings, but it was beyond the reach of fire department aerial ladders. The victims were trapped and died before fire fighters could get to them. There were no sprinklers in the penthouse and the restaurant area had combustible ceiling tiles, wall panelings, and decorations.

On May 22, 1968, two girl students at Ohio State University died of smoke inhalation from a fire on the eleventh floor of Lincoln Tower Dormitory. The fire was discovered in the living room and heavy smoke and flames developed. Fire fighters had to climb stairs to reach the eleventh floor because elevators initially could not be moved from the first floor.

On December 5, 1968, workmen were installing wood parquet flooring on the twentieth floor of an office building in Atlanta, Georgia. Unfortunately, they were using a mastic adhesive that contained an extremely flammable solvent. After they had spread this mastic in the reception area, vapors from the mastic suddenly were ignited from some unknown source. Fire quickly flashed through the three rooms killing a receptionist and trapping three workmen in the conference room. Two tried to escape from a window and fell twenty stories to their deaths. The third worker was fatally overcome in the conference room.

In the Hawthorne House apartment building in Chicago, fire occurred on the thirty-sixth floor January 24, 1969. Apparently it began when someone left a cigarette in a couch or chair in an apartment living room. It smoldered for some time, building up considerable heat which finally caused windows to crack and fall out. This allowed sufficient fresh air into the room to cause a flashover fire. The occupant of the apartment and three other persons died in subsequent fire action.

Six people died in Rault Center in New Orleans, Louisiana, on November 29, 1972. Five of them were women trapped in the beauty salon on the fifteenth floor where fire had started in a meeting room. Later, a man, trapped on an elevator died of smoke inhalation. About 100 to 150 people in a restaurant on the sixteenth floor were able to escape to the nearest stair tower and eight persons were rescued from the building roof by means of a helicopter.

On the following day, November 30th, fire occurred in the Baptist Towers Home for Senior Citizens in Atlanta, Georgia. This was an eleven-story fire-resistive apartment building designed to house the elderly. The fire started in combustible materials in a seventh floor apartment, spread through an open door to the adjacent corridor and to other apartments. Eight residents of the seventh floor and a guard died; another resident from the tenth floor died six days later. The fire department had difficulty in placing apparatus for fire control since the building was located on a sloping grade.

In New York City, a comprehensive fire prevention code was established for high-rise buildings because of the many incidents that occurred. One of the most difficult was the fire of August 5, 1970, which involved the thirty-first, thirty-second and thirty-third floors of One New York Plaza. Two building guards died in the subsequent smoke development. Less than six months later, on December 4, 1970, fire occurred on the fifth floor of a forty-seven story building on Third Avenue. Even though the fire department was able to conduct some of its operations from the ground, considerable damage spread to the sixth floor of the fire-resistive building. A major problem was evacuation of tenants above the involved areas. Three construction workers died on the fifth floor and at least twenty other persons were injured or overcome by smoke.

(United Press photo.)

In Chicago, the 110-story Sears Tower while under construction, had fire occur in an elevator shaft and four workmen were trapped and killed.

In the same city, the 100-story John Hancock Center established a new record for a high-rise fire: it started on the ninety-sixth floor and did major damage to the ninety-fifth and ninety-seventh floors also.

In Buffalo, New York, fire fighter Frank Podsiadlo plunges head first from burning building after a backdraft explosion. On the ground, Fire Fighters Ralph Backes and Don Meca prepare to break his fall. (Photo by Ron Moscati, Buffalo Courier-Express.)

Epilogue — The Challenge of Fire Control[1]

In the past twenty years the fire death toll in the United States has averaged about 12,000 per year. In the most recent year in which statistics on property loss were compiled prior to publication of this book (1974), the tabulation showed that approximately three million fires in that year had caused property damage of $4.4 billion. In the preceding five-year period, the individual fires that caused the largest loss of life included seven aircraft incidents, three ship fires or explosions, a mine disaster, two water construction projects in which methane gas was ignited by drilling operations, and a railroad tank car explosion.

As of this date the largest loss of life by fire in the United States was recorded October 9, 1871 when the Village of Peshtigo, Wisconsin burned. (*See page 87.*)

The largest loss of life in a single building fire occurred in the Iroquois Theatre December 30, 1903. (*See page 116.*)

After more than 370 years the fire problem in the United States is still a matter of major proportion, sometimes aggravated by the manufacturing, shipment and storage of new materials produced in our rapidly changing technological society, but more often by failure to apply basic principles of fire safety. The task of developing adequate defense against fires and explosion requires thorough analysis and understanding of potential fire problems; the organization and training of fire departments that can offer quick response and efficient fire fighting for large and small communities; the development of portable, mobile and fixed extinguishing equipment, designed for specific fire situations; modern communication systems and equipment, including fire detection and alarm devices; and comprehensive, modern laws and ordinances supple-

mented by inspection and enforcement. All of these are essential to the overall problems of fire control and extinguishment.

Modern Fire Departments

In the United States there are about 1,600 fully paid fire departments with approximately 185,000 personnel, and there are more than 21,000 call and volunteer fire companies having one million personnel. Small communities may have fire departments which include paid, call and volunteer personnel. The size, organization, type of operations and capabilities of fire departments vary considerably, but the normal chain of command is from the fire chief, to his subordinate chief officers, to fire company officers and then to fire fighters. Chief officers or company officers may be in charge of training, fire prevention, communications and maintenance.

The basic operational unit of a fire department is the fire company, which normally includes one or two pieces of apparatus, a company officer, and an assigned complement of five to seven fire fighters. A large fire department will have engine and ladder companies, rescue squads, perhaps two or more ambulances, maintenance and communication cars, plus cars for chief officers.

On the fireground the engine companies and ladder companies work together in teams, the former operating from pumpers and/or hose wagons, while the ladder company performs rescue, ventilation, forcible entry and other tasks.

Standard fire department pumpers are manufactured to discharge water at the rate of 500 to 1,500 gallons per minute with the pumper discharging at 150 pounds per square inch, but pumpers have been made in larger capacity, sometimes 2,000 to 2,500 gallons per minute. The famous "Super Pumper" developed for New York City during the 1960s was capable of moving 10,000 gallons per minute.

Aerial ladders are made to extend to operating heights of 65, 75, 85, and 100 feet but some American makes have been designed for 150-foot elevation.

Elevating platform apparatus may be of the articulated or telescopic type, with the platform basket raised to elevations of 65 to 150 feet depending upon the manufacturer's model.

Special kinds of apparatus are needed for aircraft crash fire situations. Generally three types of vehicles are made for this specific mission: major fire fighting vehicles with gross weights of eight or more tons, carrying special extinguishing agents; light rescue vehicles with gross weights of under four tons; and water tank vehicles. The lighter vehicles carry rescue tools and small capacity extinguishing equipment so they can get to the site quickly and start rescue operations.

Water supply for fire fighting will vary, depending upon the community and the location of the incident. Hydrant systems differ according to the size of municipality, age of the water system, number and capacity of pumping stations and location and spacing of hydrants. Static water supply, such as from ponds, streams, rivers and lakes is vital for fire fighting in many areas.

The community fire alarm system and alerting means for off-duty personnel are highly important for successful fire fighting operations. In recent years, extensive use of two-way radio has led to major improvements in fire analysis and fire fighting operations. Fire departments conduct frequent inspections of buildings and, with the help of computers and other data processing equipment, develop a very useful library of building floor plans and other illustrations that can be made available immediately for transmittal of information to the fireground. Thus, chief officers can be informed quickly of the elements of the building design and contents that are important for speedy fire control.

Extinguishing Agents

A variety of gaseous, liquid and solid material is used for fire extinguishment. Some of these agents are applied through hose and nozzles; others by portable, hand-carried fire extinguishers; by wheeled extinguishers; or by fixed systems. Each extinguishing agent has certain characteristics and capabilities, the principal limitations being the amount of agent available for a particular fire incident, and the efficiency of application.

Plain water is the most commonly used extinguishing agent because it can be made available in large quantities and is suitable for most fire situations. Water accomplishes extinguishment by cooling the surface of the burning material; by smothering (when the water turns to steam which can displace the air); by emulsification, when certain liquids are agitated together; and by dilution of certain flammable materials that are soluble in water. Water is usually applied by nozzles and hose lines in straight or spray streams, or by sprinkler systems.

Certain chemicals can be added to water to change its extinguishing ability for specific reasons. Antifreeze may be added to lower the freezing point of water in cold climates. Calcium chloride solutions are useful for this purpose. Water can be made "wetter" by the addition of a chemical that decreases the surface tension, thus increasing the penetrating and spreading qualities of the water. Other chemicals can be added to make water slippery, more viscous, to improve its smothering and adhering characteristics. This is common in forest fire fighting where water is frequently dropped from aircraft as a gel.

Foam

A number of foaming agents are used in fire fighting. The foam produced for this purpose is a mass of gas-filled bubbles which, when applied properly, produce an air-excluding, cooling, continuous layer of vapor-sealing, water-bearing material. This floats on top of flammable or combustible liquids, or clings to solid material, to reduce the heat and smother the fire by excluding oxygen. Foam may be applied

In 1967, Astronauts Virgil E. Grissom, Edward H. White, II, and Roger B. Chaffee, members of the Apollo flight crew, died during a launch simulation test at Cape Kennedy, Florida. They were in an atmosphere of pressurized oxygen, practicing routines for space operations, when ignition occurred. They burned to death before they could reach the escape hatch. The square hatch at center in this photo, was used for servicing craft components. (UPI Telephoto.)

In 1973, a tank truck carrying flammable liquid collided with a passenger car beneath an overpass in Glen Burnie, Maryland. Responding volunteer fire departments used high expansion foam, covering the roadway to a depth of about five feet and enveloping the two vehicles. (Photo by Joseph Gruver.)

through hose lines and nozzles, by portable or wheeled extinguisher, by special applicators, or from fixed systems. The kind of foam most suitable for a given fire situation will depend upon the type of fire, the kind of material involved, and other factors, such as strength of air movement, flowing of fuel and impediments to the foam pattern.

Foam made from protein type concentrates usually is proportioned to a concentration of three or six percent by volume, and produces dense, viscous foam of high stability and high heat resistance.

Fluoroprotein foaming agents usually feature the addition of a fluorine compound as a surface active agent that confers a "fuel shedding" property to the foam. Thus, fluoroprotein foams are more effective on certain petroleum or hydrocarbon fuel fires or for subsurface injection of the foam in tank fire fighting.

Low temperature foaming agents are used in cold regions where temperatures may be as low as twenty degrees below zero Fahrenheit.

Aqueous film-forming foam agents (AFFF) also contain fluorinated hydrocarbons and they create an air foam similar to those produced by protein-based materials. They also develop a layer of water solution under the surface foam, thus forming a film floating on top of hydrocarbon fuel surfaces to cool the fuel and suppress flammable vapors.

"Alcohol-type" foaming agents are used for fires involving alcohol, enamel and lacquer thinners, acetone, isopropyl ether and other fuels that are water soluble, water miscible, or "polar solvents."

"High expansion" foam can be used for extinguishment of fire in combustible materials such as wood, or flammable liquids. It is particularly suitable as a flooding agent for a room or other confined space. This kind of foam expands in ratios of 100 to 1,000 to 1 and, when used in sufficient volume with the good confinement, can achieve quick extinguishment of fire. It is particularly useful in flooding basement room areas or holds of ships where fires are difficult to reach.

Carbon Dioxide

Carbon dioxide (called CO_2) is used for the extinguishment of fires in flammable liquids, gases, and electrically energized equipment. Carbon dioxide is stored under pressure as a liquid and is discharged into the fire location as a gas. When applied in adequate quantity it will envelop the burning material and dilute the oxygen to a concentration that cannot support combustion. As the gas is discharged it expands rapidly, producing a refrigerating effect that converts part of the carbon dioxide into snow that has a temperature of minus 110 degrees F. As this snow turns into gas it absorbs heat from the burning material in the surrounding atmosphere.

Carbon dioxide is applied by portable fire extinguishers, by hose lines attached to pressurized tanks, by special nozzles operating from tanks on fire apparatus, or by fixed extinguishing systems. It does not work well on hot surfaces and embers, or materials that contain their own oxygen supply, and reactive chemicals.

Halogenated Extinguishing Agents

Halogenated (halon) extinguishing agents are hydrocarbons in which one or more hydrogen atoms have been replaced by halogen atoms. The common elements are fluorine, chlorine, bromine, and iodine. The halogenated agents are very suitable for use on electrical fires because they have low electrical conductivity. They are easily contained or stored as a gas or liquid, and leave little corrosive or abrasive residue after use on a fire. In fixed systems, halogenated agents are usually applied as a gas; in portable application they are applied as a liquefied gas or vaporizing liquid. These agents have been very effective for flame extinguishment and explosion suppression.

Dry Chemical

Another common extinguishing agent is dry chemical. Mixtures of dry powdered chemicals have proven effective for smothering and extinguishing fire involving flammable liquids and certain chemicals. There are five basic varieties: "regular" and "ordinary" dry chemical, accepted for use on flammable liquid and electrical fire;

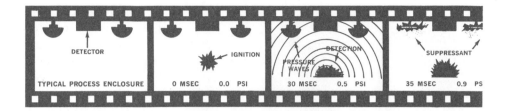

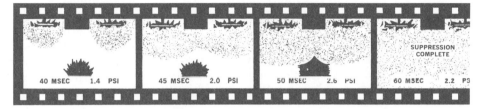

Sketch of how an explosion suppression system operates in milliseconds to stop ignition and pressure waves.

"multipurpose" dry chemical, usable on ordinary, flammable liquid and electrical fires; and special dry powders and compounds used on combustible metal fires.

In years past, ordinary dry chemical was a mixture of borax and sodium bicarbonate. In 1959 sodium bicarbonate was modified to render it compatible with protein-based low expansion foam to permit a "dual agent" attack of dry chemical and a foaming agent. A year later two other dry chemicals were developed, one with ammonium phosphate base and the other with potassium bicarbonate base. These were succeeded by another agent with a potassium chloride base. Each of these compounds has specific capabilities and is designed for certain applications, but they are used primarily on flammable liquid fires and on fires involving live electrical equipment. Smothering, cooling, and radiation shielding contribute to the extinguishing efficiency of dry chemicals. Studies also indicate that a chain-breaking reaction interrupts the flame when this extinguishing agent is applied.

Combustible Metal Extinguishing Agents

Certain types of metals burn when heated to high temperature by some outside source, or from contact with moisture, or reaction with other material. Extinguishing these fires can be hazardous because there can be a steam explosion, explosive reaction with one of the common extinguishing agents, dangerous radiation from certain nuclear material, and combustible gases or toxic products of combustion. In manufacturing or storage, some of these metals are in small particles such as chips, fines, turnings, or powder. In these forms they can be ignited quite easily. Magnesium,

titanium, uranium, zirconium, sodium, potassium, lithium and aluminum are combustible materials.

A number of extinguishing agents have been developed for such metal fires. These are in powder, flux or liquid form and may be applied by shovel or discharged from portable fire extinguishers.

Water can be used on some combustible fires but it must be applied cautiously and in quantity appropriate for the particular metal.

The Importance of Automatic Sprinklers[2]

Ideally, to extinguish a fire, it is essential to apply as much extinguishing agent as necessary in the quickest operation with the best distribution, so the fire is cooled and/or smothered in the least possible time. Automatic extinguishing systems accomplish each of these operations. Sprinkler systems, using water as the cheapest and most readily available extinguishing agent, are the most well-known, and most easily recognized of these systems. They are designed to distribute water quickly and efficiently as soon as sprinkler heads are activiated by a fire temperature.

Essentially, an automatic sprinkler system consists of a basic source of water supply, such as an elevated tank or municipal water system; one or more stationary pumps to increase pressure when needed; a network of piping to convey water to inside or outside areas of protection; valves, operated automatically or manually; an appropriate number of sprinkler heads, designed to function in a certain pattern of discharge, and spaced according to the anticipated fire protection need; plus mechanical, hydraulic, and electrically operated components that make that system efficient.

At left is the Parmelee sprinkler, the first successful automatic sprinkler head, manufactured and tested in 1874. At center is a modern upright sprinkler. At right is a concept of how a future efficient sprinkler head might look.

Each sprinkler "head" usually has a half-inch orifice and is designed to distribute water over an area of slightly less than one hundred square feet, at a rate of one-tenth to two-tenths gallons per square foot per minute. The pattern of discharge of the sprinkler head should be engineered for the particular hazard. To the layman, sprinkler systems appear complicated, or perhaps aesthetically obtrusive, in rooms or buildings whose ceilings (from the designer's point of view) should be "uninterrupted." It must be recognized, however, that automatic sprinkler systems have attained an outstanding record of performance in the past century; in most years their overall performance in fires has been carefully recorded and has generally exceeded design anticipations. Modern sprinkler heads can be aesthetically pleasing and unobtrusive as well as functionally efficient.

The concept of delivering water by a system of piping to control a structural fire probably originated hundreds of years ago, but the first practical installation in the United States was in a textile mill in Lowell, Massachusetts about 1852. James B. Francis, hydraulic engineer of the mill of the Proprietors of Locks and Canals on the Merrimac River, designed a system of perforated pipes to which water could be supplied for fire extinguishment. The system had to be turned on manually and distribution through the perforations was crude; but for about thirty years it was an accepted practice in the New England cotton mills. It should be noted that, until suitable pipe was available for carrying water at something more than low pressures, adequate sprinkler systems could not be achieved. In those years, some municipalities still were using wooden pipe for water mains; lead and wrought iron served for metal pipe.

Henry S. Parmelee was the first to invent a sprinkler which would discharge water automatically when fire occurred. He created a sprinkler head that would operate when a small string or cord burned through; but he quickly changed this when his piping contractor suggested the use of a fusible solder. Parmelee's first patent was for a device with a perforated rosette that would begin distributing water as soon as the solder component melted. He received a patent for this device in 1874, and it quickly caught the attention of industry owners and insurance interests. Parmelee made arrangements with the Providence Steam and Gas Pipe Company for marketing and installing his sprinkler systems. Later, Frederick Grinnell of that company redesigned the sprinkler fitting to make it more sensitive to fire. From that time on, sprinkler systems underwent constant and frequent change.

Between 1872 and 1914 some 450 patents were granted by the U. S. Patent Office for components of the precision mechanisms that make up these systems. Despite the crude limitations of the first devices, some 200,000 Parmelee sprinklers were installed between 1875 and 1882, mostly in New England mills. From then on, as long as water supply was available, there was a steady increase in the use of sprinklers for fire protection.

These systems have had an exceptional record of performance. Loss of life by fire in sprinklered buildings has been minimal, usually a matter of individual fire death, never a large group of victims. The limitation of a sprinkler system is in the water supply; if an explosion breaks piping, or otherwise prevents water from reaching the fire, or if a main control valve is closed, the system cannot function. These things have happened and tremendous fire losses have resulted; on the other hand, many thousands of injuries and fatalities and perhaps billions of dollars of property loss have been avoided because sprinklers have controlled or extinguished fires.

Until 1952 water discharged from sprinklers was regulated in a somewhat coarse pattern by the design of the sprinkler head; little attention was given to the optimum efficiency of the water spray. Then in 1952 and 1953 the spray sprinkler head was developed and it provided a more efficient distribution of fine water spray which gave maximum cooling capacity at less water flow. Five years later this became accepted as the standard discharge pattern. The next major advance came in the 1970s when the on-off recycling system, and then the on-off sprinkler heads were developed. With these, water is applied according to fire demand; the system or sprinkler head is actuated by heat, discharges, and when the temperature drops, the head, or system, closes down. It operates again if heat conditions indicate the need.

Present research is directed toward making automatic sprinkler systems completely self-inspecting

and self-supervising, requiring little or no human attention. Computers and other automatic indicators will maintain constant watch on all functioning parts of the system and will actuate when failure or other trouble signal demands. Other research is aimed at determining the optimum shape of water discharge, size of water drops, and the best way for penetrating the rising heat column with appropriate water volume and pressure. The use of water additives for improving the smothering capabilities of extinguishment is also a matter of research.

Special Extinguishing Systems

Many manufacturing processes require special systems that operate quickly and efficiently to control fires and explosions. Each must be designed specifically for the anticipated hazard and all feature highly sensitive detection and actuation equipment to meet requirements of instant operation. Such installations include systems using water, carbon dioxide, and dry chemical, foam, or a combination of these agents; or steam smothering systems using combustion gases from some process in an industrial plant. Usually, such a system includes a storage container of the extinguishing agent(s), piping and nozzles for distribution of the agent detection units and system actuation equipment. All are arranged according to specific design and requirements of fire protection standards.

Ultra-high-speed water spray systems deliver huge quantities of water in spray discharge within thousandths of a second (millisecond). For example, in one group of hyperbaric test chambers, the time elapse from fire ignition to delivery of water at designed nozzle pressure was 96 milliseconds. Another system, designed for protecting a solid propellant installation, operated in about 240 milliseconds. The discharge of such a system is intended to overpower the flame or pressure wave of explosion that would otherwise occur.

Detectors and Signaling Systems

During the past 200 years protective signaling systems have been used in fire protection to indicate when and where a fire occurs, or when a sprinkler system operates, and to transmit this information to the local fire department or a central signaling station. Within the past few years systems have been developed specifically for the protection of single family dwellings, apartment buildings, churches, nursing homes, hospitals and other structures in which fire can endanger human life.

These automatic systems may use heat detectors, sensitive to abnormally high temperature, or to rate of temperature rise; smoke detectors, sensitive to

When the nuclear powered attack aircraft carrier *U. S. S. Enterprise* was maneuvering seventy-five miles south of Pearl Harbor in 1969 a huge fire started on the after deck. Destroyer *U. S. S. Rogers* provided fire fighting support. (Official U.S. Navy photos.)

In the Vietnam war, fire was used frequently to burn villages, but it was also dropped from aircraft in the form of petroleum jelly which ignited vegetation and caused severe burns on humans. At sea, fire was a constant threat. This photo shows the attack aircraft carrier *U. S. S. Oriskany* with its hangar bay involved. Forty-four crew members died as a result of this blaze.

In the Gulf of Tonkin off Vietnam on July 29, 1967 the aircraft carrier *U. S. S. Forrestal* suffered a series of blasts as heavily armed, fueled aircraft were being prepared for launch on combat trips. One hundred and thirty-two lives were lost, sixty-two men were injured and three were missing after the incident.

Members of the Toronto, Ontario Fire Department rescue a 3-year-old boy from the wreckage of his home which was devastated in an explosion. Saw in background was used to cut heavy floor joists and timber. (Toronto Fire Department photo.)

invisible or visible particles of combustion; or flame detectors, sensitive to the infrared, ultra-violet, or visible radiation produced by a fire. Detection units must be installed according to fire protection standards. Some detectors are battery operated; others are supplied by the building's electrical system.

A horn or other alerting device may be a part of the detector or may transmit a signal to an alerting device in some other part of the building, such as a watchroom in a manufacturing plant; or to a central station in some part of the community, from which the signal agency personnel retransmits the alarm to the fire department, or takes other necessary action; or to the local fire department which will respond to handle the situation. It is most important for the signal to be transmitted promptly to the fire department, particularly for fire incidents in dwellings. In some localities this type of installation can be achieved through the telephone system. These home fire warning systems are designed primarily for life protection, not for the protection of property.

Fire Prevention

Even though the means of fire extinguishment are highly efficient and are being improved continually, the most important defense against fire is fire prevention. This must be accomplished in every portion of our society: by the general public, by fire departments, by responsible persons in business and industry, and by legislative action on local, county, state and national levels. The sampling of incidents illustrated and described in this book should provide convincing evidence that our entire nation must develop an intelligent system of continual fire prevention if we are to minimize the tragic and disastrous losses that can occur.

The general public needs to be informed continually of the basic fire hazards and of specific dangers within communities or areas. In addition to the common sense practices of fire safety, people must know how to escape from a burning building, how to react to fire situations, the fire dangers of certain buildings, and how to summon the local fire department quickly. Seasonal warnings are necessary when there is outdoor fire danger, or special indoor hazards such as the crowding of restaurants and stores and the large accumulation of combustible materials during holiday seasons.

The public Fire Service has a major task in carrying out its inspection of buildings and public information programs. It needs to present instructions for schools, local civic organizations, industry, institutions, and professional and commercial groups. At the same time it must maintain its apparatus, equipment and fire fighting capabilities.

Property owners and managers must be informed of the meaning of fire protection standards and legislation which affect the construction and operation of their buildings. Similarly, elected officials in municipalities, counties and states must be aware of the need for evaluating legislation pertaining to fire protection and prevention. There are national codes and standards which can be adopted in the law or otherwise utilized to update legislation and enforcement of fire prevention requirements.

Fire fighter Arthur J. Johnson of Engine Company 7 in Erie, Pennsylvania, tries desperately to revive a nine-month-old boy who was suffocated in a fire, started by children playing with matches. The child did not revive. (Photo by Joseph A. Comstock, Erie Times-News.)

Appendix

Fire Museums/Collections — United States and Canada

Listing by Harold S. Walker, Marblehead, Massachusetts

The Phoenix Museum
203 South Clairborne Street
Mobile, Alabama

Hall of Flame
National Historical Fire Foundation
Scottsdale, Arizona

Firemen's Fund Insurance Company
San Francisco, California

Knott's Berry Farm
Buena Park, California

Old Plaza Fire House
Los Angeles, California

State Park
Columbia, California

Sutter Museum
Sacramento, California

Connecticut Fire Museum
Warehouse Point, Connecticut

Delaware State Museum
Dover, Delaware

Smithsonian Institution
Washington, DC

The Carriage Cavalcade
Silver Springs, Florida

Jacksonville Fire Department Museum
12 Catherine Street
Jacksonville, Florida

Museum of Science and Industry
Chicago, Illinois

Louisiana State Museum
New Orleans, Louisiana

Fire Museum of Maryland, Inc.
1301 York Road
Lutherville, Maryland

Antique Auto Museum
15 Newton Street
Brookline, Massachusetts

Berkshire Museum
Pittsfield, Massachusetts

Edaville Railroad and Museum
South Carver, Massachusetts

Essex Institute
Salem, Massachusetts

New England Fire and History Museum
Brewster, Massachusetts

Old Sturbridge Village
Sturbridge, Massachusetts

Society for the Preservation of New England Antiquities
Norwell, Massachusetts

Townsend Fire Department Museum
Townsend, Massachusetts

Henry Ford Museum
Dearborn, Michigan

Kalamazoo Public Museum
Kalamazoo, Michigan

Austin Firefighting Museum
Austin, Minnesota

Missouri Historical Society
Forest Park
St. Louis, Missouri

Minden Pioneer Village
Minden, Nebraska

Last Frontier Village
Las Vegas, Nevada

Nevada State Museum
Carson City, Nevada

Manchester Historic Association
Manchester, New Hampshire

James E. Lesnick Fire Museum
217 Willow Avenue
Piscataway Township, New Jersey

Newark Fire Museum
41–49 Washington Street
Newark, New Jersey

Trenton's Firemen's Museum
244 Perry Street
Trenton, New Jersey

American Museum of Fire Fighting
Firemen's Home
Hudson, New York

Home Insurance Company
59 Maiden Lane
New York, New York

Long Island Automotive Museum
Southampton, New York

Long Island Old Fire Engine Collection
Baldwin, Long Island, New York

Museum of the City of New York
8th Avenue and 103rd Street
New York, New York

New York City Firemen's Museum
104 Duane Street
New York City, New York

Signal 7–7 Museum
326 Livingston Street
Brooklyn, New York

Smith's Cove
Old Museum Village
Monroe, New York

Ye Olde Fire Station Museum
8662 Cicero-Brewster Road
Cicero, New York

Allen County Historical Museum
Lima, Ohio

Cincinnati Fire Department Museum
9th and Broadway Fire Station
Cincinnati, Ohio

Ohio State Museum
Columbus, Ohio

Oklahoma State Firemen's Museum, Inc.
2716 N. E. 50th Street
Oklahoma City, Oklahoma

Buda County Historical Society
Doylestown, Pennsylvania

Doylestown Museum of Historical Vehicles
Doylestown, Pennsylvania

The Franklin Institute
Philadelphia, Pennsylvania

Hershey Museum
Hershey, Pennsylvania

Insurance Company of North America
1800 Arch Street
Philadelphia, Pennsylvania

Lincoln Fire Station
Bethlehem, Pennsylvania

Old Economy Museum (Harmony Society)
Ambridge, Pennsylvania

Philadelphia Fire Department Museum
149 North 2nd Street
Philadelphia, Pennsylvania

Charleston Museum
Charleston, South Carolina

Memphis Fire Department Museum
2 Avery Street
Memphis, Tennessee

Dallas Firefighters Museum
3801 Parry
Dallas, Texas

San Antonio Museum Association
San Antonio, Texas

Shelburne Museum
Shelburne, Utah

Friendship Fire Engine Company Museum
107 South Alfred Street
Alexandria, Virginia

Gamon Museum & Wagons
Mabton, Washington

Washington State Fire Service Historical Museum
Seattle, Washington

Milwaukee County Historical Society
Milwaukee, Wisconsin

CANADA

Chateau de Ramsay
Montreal, P. Q., Canada

Fire Fighters Museum of Nova Scotia
Yarmouth, N. S., Canada

Toronto Fire Department Museum
Eastern Avenue
Toronto, Ont., Canada

Steam Fire Engine Builders in U.S.A. Who Built More Than Five Engines
Listing by Harold S. Walker, Marblehead, Massachusetts

Latta and Ahrens types

Latta & Shawk, Cincinnati	1852–1854	9
A. B. & E. Latta, Cincinnati	1855–1866	11
Lane & Bodley, Cincinnati	1862–1868	18
C. Ahrens Company & Ahrens Manufacturing Company	1868–1890	434

(Plus 73 Ahrens Engines Built by the American Fire Engine Company after the consolidation.)

Silsby types

Silsby, Mynderse & Company	1856–1860	23	
H. C. Silsby, Joland Works	1861–1870	150	801
Silsby Manufacturing Co.	1870–1890	560	
American Fire Engine Co.	1890–1900	68	
Clapp & Jones Manufacturing Company		391	460
American Fire Engine Co.		69	
Button & Blake			198
L. Button & Son			
Amoskeag Manufacturing Co.	1859–1879	545	
Manchester Locomotive Works	1879–1902	225	953
American Locomotive Co.	1902–1908	169	
International Power Co.	1908–1913	14	

Lee & Larned, New York   No Data
Abel Shawk, Cincinnati
James Smith, New York
Joseph Banks, New York
Murray & Hazelhurst
Poole & Hunt, Baltimore

Shepherd Iron Works, Buffalo	No Data
Reaney & Neafie } Philadelphia	
Neafie & Levy	
G. J. & J. L. Chapman, Philadelphia	
Portland Company, J. B. Johnson, Designer, Portland, Maine	
Ettenger & Edmond, Richmond, Virginia	
William Jeffers, Pawtucket, Rhode Island	63 } 68
Skidmore & Morse, Bridgeport, Conn. (Successor to Jeffers)	5
Ives & Son } Baltimore	12
John A. Ives & Bro.	
Cole Bros., Pawtucket, Rhode Island	47
Gould Machine Company }	No data
E. S. Nichols & Company	
Hunneman & Company	30
Allerton Iron Works	12
Jucket & Freeman, Boston	
Union Iron Works, Fitchburg, Mass. } All "Jucket" Engines	19
Allen Fire Department Supply Company Providence, Rhode Island	
John H. Dennison }	11
Dennison Manufacturing Company	
Campbell & Rickards	7
LaFrance Fire Engine Company, Elmira, New York 350 Piston type, 60 rotary type	410

About 15 or 20 also built by American LaFrance Company after consolidation.

(*continued next page*)

Thomas Manning, Jr. & Company, Cleveland	30
Mansfield Machine Company, Mansfield, Ohio	30
Waterous Engine Works, St. Paul, Minnesota	75
Ahrens Fire Engine Company Cincinnati, Ohio — Continental S.F.E.	
Ahrens-Fox Fire Engine Company 1906–1915	203

W. S. Nott Company, Minneapolis, Minnesota	150
American LaFrance Fire Engine Company 1903–1915	350
American Fire Engine Company Seneca Falls and Cincinnati 1890–1903	
	Est. 350–500

Sources of Information

In the following pages, the Bibliography refers to specific publications and articles which were used as source material for this book. It is important, also, to acknowledge assistance received by the author from some of the principal organizations and agencies that provided illustrations and information on fire. Readers who wish to make more extensive study of any subjects mentioned in this book will find these organizations and agencies of great help.

From Chapter 5 to the end of the book, the principal source of information was the National Fire Protection Association, 470 Atlantic Avenue, Boston, Massachusetts 02210. The NFPA is a scientific and educational membership organization concerned with the causes, prevention and control of destructive fire. It has a diverse membership of more than 32,000 and is recognized world-wide as the most authoritative fire protection organization. The Association has thousands of fire reports that have been developed since the NFPA was organized in 1896. Most of the fires summarized in this book have been digested from those reports and sources of the others are referenced in the Bibliography.

The NFPA has more than 150 technical committees each having a balanced representation of interests, working continuously for the development and improvement of fire protection standards. More than 2,400 well-qualified members of these committees serve voluntarily without compensation to solve the problems of fire protection by application of technical knowledge.

The Association also provides information exchange on significant fires and fire-related developments; holds two major meetings annually; publishes monthly and bimonthly periodicals; and develops a large quantity of books and pamphlets dealing with all aspects of fire. *The Fire Protection Handbook*, recognized as the most authoritative volume in the field, was published in a fourteenth edition in 1976.

The NFPA has a variety of data collection, public information, analysis and research programs, participates in activities of the International Standards Organization, a unit of the United Nations Organization in Geneva, Switzerland, and works with the American National Standards Institute and several consensus standard and testing organizations in the United States.

The Library of Congress in Washington, D. C. has a large collection of books, illustrations and original documents on a great many subjects and it was a very useful source for pictures of early fires and ships in wartime battles.

The National Archives in Washington has an excellent reference library and a large amount of microfilmed documents pertinent to every decade of our nation's history. There is a wealth of information covering the years prior to and during the Revolutionary War, some excellent volumes on the Civil War and subsequent military and naval engagements, and a vast quantity of resource material on thousands of subjects. In addition, there is a well-informed staff of professional archivists and other specialists who are of great assistance in tracing source materials.

The Smithsonian Institution in Washington has a deserved reputation as one of the leading museums in the world. It has an extensive collection of black and white and full-color prints and negatives showing fire scenes, fire apparatus, old-time equipment, famous personalities and Currier and Ives prints. In addition, there are many excellent items of fire apparatus and equipment, restored and preserved in full authenticity for public display. The curators of the various departments of the Institution are obliging sources of information on historical matters.

The Curator Branch of the Navy Museum in the Washington Navy Yard has a large photographic section with illustrations of many fires and sea battles, and the U. S. Naval Photographic Center in Anacostia, D. C., has a comprehensive library of "still" pictures showing fires and other incidents on Navy property.

The Photographic Center of the Pentagon, the U. S. Patent Office in Crystal City, Virginia, and the U. S. Energy Research and Development Administration (atomic weapons) in Washington are other useful sources of information.

The U.S. Forest Service, Washington, D.C., and the Department of Interior, Mining Enforcement and Safety Administration, Pittsburgh, Pennsylvania provide much helpful data.

Bibliography

Chapter I

[1] *Boston Records*, 1634 to 1660 and *Book of Possessions* (1884 edition).

[2] Lovejoy, David S. *The Glorious Revolution in America*. Harper and Row, New York, New York.

[3] Dunshee, Kenneth Holcomb. *As You Pass By*. Hastings House, New York, New York. 1952.

[4] Governor William Bradford's History of Plymouth Plantation, Book II, pp 182–183. Printed from the original manuscript under the direction of the Secretary of the Commonwealth of Massachusetts by order of the General Court. 1898.

[5] Dunshee. *As You Pass By*.

[6] History of the Boston, Massachusetts Fire Department, 1888.

[7] Bragdon, Henry W. and McCutchen, Samuel P. *History of a Free People*. The MacMillan Company, New York, New York. 1967.

[8] History of the Boston, Massachusetts Fire Department, 1888.

[9] *Ibid.*

[10] *Ibid.*

[11] State Fire Marshal James C. Robertson, Baltimore, Maryland. Personal letter.

[12] History of the Boston, Massachusetts Fire Department.

[13] History of the Charleston, South Carolina Fire Department. Mimeograph.

[14] History of the Richmond, Virginia Fire Department. Mimeograph.

[15] White, Captain Charles E. *The Providence Firemen*. E. L. Freeman and Son, State Printers, Providence, Rhode Island. 1886.

[16] Dunshee. *As You Pass By*.

[17] Gilbert, K. R. *Fire Engines and Other Fire-Fighting Appliances*. Her Majesty's Stationery, London, England.

[18] Letter from Savannah, Georgia Fire Department, 1975.

[19] Letter from Baltimore, Maryland Fire Department, 1975.

[20] Historical report, Wilmington, Delaware Bureau of Fire, 1975. Mimeographed.

[21] Letter from Newark, New Jersey Historical Fire Association, 1975.

[22] Historical summary, Louisville, Kentucky Fire Department. Mimeographed.

[23] History of the Washington, D. C. Fire Department. 1936.

[24] Gilbert. *Fire Engines, etc.*

[25] History of the Boston Fire Department.

[26] Gilbert. *Fire Engines, etc.*

[27] *Ibid.*

[28] Centennial Report, Philadelphia, Pennsylvania Fire Department, 1971.

[29] NFPA Handbook of Fire Protection, 12th edition.

[30] Lovejoy, David S. *The Glorious Revolution in America*.

[31] *Lossing's Field Book of the Revolution*. U.S. National Archives.

[32] Cummings, William P. and Rankin, Hugh. *The Fate of a Nation*. Phaidon Press, Limited, London, England. 1975.

[33] *Lossing's Field Book of the Revolution*. U. S. National Archives.

[34] White, Captain Charles E. *The Providence Firemen*.

[35] Weigley, Russell G. *The American Way of War*. The MacMillan Company, Inc., New York, New York. 1973.

[36] Miller, Nathan. *Sea of Glory*. David McKay Company, Inc., New York, New York. 1974.

[37] Limpus, Lowell M. *History of the New York Fire Department*. E. P. Dutton Company, Inc., New York, New York. 1940.

[38] Weigley, Russell G. *The American Way of War*.

[39] Sheldon, George W. *Story of the Volunteer Fire Department of the City of New York*. Harper and Brothers, New York, New York. 1882.

[40] Weigley, Russell G. *The American Way of War*.

[41] *Lossing's Field Book of the Revolution*. U.S. National Archives.

Fire Marks

[1] NFPA *Volunteer Firemen* Magazine, November, 1943. Article by the Insurance Company of North America.

Chapter II

[1] History of the Saint Louis, Missouri Fire Department. Mimeographed.

[2] History of the Washington, D. C. Fire Department. 1936.

[3] *Ibid.*

[4] Limpus, Lowell M. *History of the New York Fire Department*. E. P. Dutton Company, Inc., New York, New York. 1945.

[5] *Ibid.*

[6] Andrews, Ralph W. *Historic Fires of the West*. 1966.

[7] History of the Saint Louis, Missouri Fire Department.

[8] From a talk by Allen L. Cobb, Eastman Kodak Company. NFPA *Firemen Magazine*, December, 1954.

[9] Dunshee, Kenneth Holcomb. *Engine! Engine!* Published by Harold Vincent Smith for the Home Insurance Company, New York, New York. 1939.

[10] Centennial Report, Philadelphia, Pennsylvania Fire Department, 1971.

[11] Letter from Charleston, South Carolina Fire Department.

[12] Horsman, Reginald. *The War of 1812*. Alfred A. Knopf, New York, New York, 1969.

[13] *Ibid.*

[14] *Ibid.*

[15] History of the Washington, D. C. Fire Department. 1936.

[16] NFPA *Volunteer Firemen* Magazine, April, 1945.

[17] Centennial Report, Philadelphia, Pennsylvania Fire Department.

[18] History of the Los Angeles, California Fire Department. Mimeographed.

[19] History of the Cincinnati, Ohio Fire Department. Mimeographed.

[20] Letter from Milwaukee, Wisconsin Fire Department.

[21] History of the Saint Louis, Missouri Fire Department.

[22] Cummings, William P. and Rankin, Hugh. *The Fate of a Nation*. Phaidon Press, Limited, London, England. 1975.

Hydrants

[1] NFPA *Volunteer Firemen* Magazine, December, 1944.

Chapter III

[1] *The War of the Rebellion*: A Compilation of the Official Records of Union and Confederate Armies. 1902. U. S. National Archives.

[2] *Ibid.*

[3] Catton, Bruce. *A Stillness at Appomattox*. Doubleday and Company, Garden City, New York. 1953.

[4] *Ibid.*

[5] Weigley, Russell F. *The American Way of War*. MacMillan Publishing Company, Inc., New York, New York. 1973.

[6] *Harper's Weekly*, July, 1865.

[7] Historical Report of the Charleston, South Carolina Fire Department. Mimeographed.

[8] *Harper's Encyclopedia of U. S. History*, Harper and Brothers, New York, New York. 1902.

[9] *Harper's Weekly*, July, 1965.

[10] Historical Report of the Richmond, Virginia Fire Department. Mimeographed.

[11] Miller, Francis. *The Photographic History of the Civil War*. Castle Books, New York, New York. 1957.

[12] *Ibid.*

[13] *Ibid.*

[14] Sheldon, George W. *History of the Volunteer Fire Department of the City of New York*. Harper and Brothers, New York, New York. 1882.

[15] History of the Washington, D. C. Fire Department.

[16] *Harper's Weekly*, July, 1865.

[17] History of the Washington, D. C. Fire Department.

[18] Headley, Joel Tyler. *The Great Riots of New York, 1712–1873*. Dover Publications, Inc., New York, New York. 1971.

[19] *Ibid.*

[20] Limpus, Lowell M. *History of the New York Fire Department*. E. P. Dutton, Company, Inc., New York, New York. 1945.

Alarms of Fire

[1] White, Captain Charles E. *The Providence Firemen*. E. L. Freeman and Son, State Printers, Providence, Rhode Island. 1886.

[2] History of the Charleston, South Carolina Fire Department. Mimeographed.

[3] History of the Cincinnati, Ohio Fire Department. Mimeographed.

[4] NFPA *Firemen Magazine*, December, 1958.

[5] History of the Los Angeles, California Fire Department. Mimeographed.

[6] Letter from the Baltimore, Maryland Fire Department.

[7] History of the Fargo, North Dakota Fire Department.

[8] History of the Indianapolis, Indiana Fire Department. 1974.

[9] History of the Phoenix, Arizona Fire Department.

[10] History of the Newark, New Jersey Fire Department.

[11] Kimball, Warren Y. *Fire Service Communications for Fire Attack*. NFPA, Boston, Massachusetts. 1972.

The Self-Propellers

[1] Article by Harold S. Walker. NFPA *Firemen Magazine*, December, 1958.

Chapter IV

[1] Brooklyn Theatre fire, page 94.

[2] Bragdon and McCutcheon, *History of a Free People*. The MacMillan Company, New York, New York. 1967.

[3] History of the Washington, D. C. Fire Department.

[4] History of the New York Fire Department.

[5] NFPA *Firemen Magazine*, March, 1955. Reprinted from bulletin of Chautaqua, New York Fire Service.

[6] NFPA *Fire Journal*, Volume 71, No. 5.

[7] Goodspeed, Reverend E. J. *Great Fires in Chicago and the West*. H. S. Goodspeed and Company, New York, New York. 1871.

[8] *Harper's Weekly*, December, 1872.

[9] NFPA *Fire Journal*, Volume 71, No. 5.

[10] NFPA *Firemen Magazine*, October, 1961.

[11] *Ibid.*

[12] NFPA *Firemen Magazine*, January, 1943.

[13] NFPA *Firemen Magazine*, November, 1961.

[14] History of the New York Fire Department.

[15] *Ibid.*

[16] History of the Newark Fire Department.

[17] History of the Los Angeles Fire Department.

[18] History of the Cincinnati Fire Department.

[19] Letter from the San Antonio Fire Department.

[20] Letter from the Portland, Oregon Bureau of Fire.

[21] Letter from the Dallas, Texas Fire Department.

[22] Letter from the San Francisco, California Fire Department.

[23] History of the Honolulu, Hawaii Fire Department.

[24] *Ibid.*

[25] Letter from the Seagrave Company.

[26] Letter from the Howe Fire Apparatus Company.

[27] Letter from the Waterous Company.

Era of Horses

[1] History of the New York Fire Department.

[2] History of the Los Angeles, California Fire Department.

[3] *100 Years of Glory.* District of Columbia Fire Department. 1971.
[4] History of the Honolulu, Hawaii Fire Department.
[5] Letter from the Phoenix, Arizona Fire Department.
[6] History of the Salt Lake City, Utah Fire Department.
[7] History of the Cincinnati, Ohio Fire Department.

Steam Fire Engines

[1] Photos and information from collection of Harold S. Walker, Marblehead, Massachusetts.

Chapter V

[1] NFPA *Quarterly* Volume 5, No. 1.
[2] NFPA Sectional Bulletin No. 54 on Iroquois Theatre fire.
[3] Laboratories Data, March, 1927. Article by Alvin F. Harlow in *Collier's Magazine.*
[4] NFPA *Firemen Magazine* February, 1964. See also NFPA Special Report F1.
[5] NFPA *Quarterly* July, 1908 Volume 2, No. 1.
[6] Editorial, NFPA *Quarterly* Volume 24, No. 1.
[7] NFPA Special Report F8.
[8] NFPA *Quarterly* Volume 4, No. 3.
[9] Nadem, Corinne J. *The Triangle Shirtwaist Fire.* Franklin Watts, Inc., New York, New York. 1971
[10] Editorial NFPA *Quarterly* Volume 4, No. 4.
[11] NFPA *Quarterly* January, 1960.
[12] NFPA *Quarterly* Volume 4, No. 3.
[13] NFPA *Quarterly* Volume 5, No. 4.
[14] NFPA *Quarterly* Volume 10, No. 2.
[15] NFPA *Quarterly* Volume 22, No. 3.
[16] *Ibid.*
[17] *Ibid.*
[18] NFPA *Quarterly* Volume 13, No. 3.
[19] NFPA *Quarterly* Volume 18, No. 3.
[20] NFPA *Quarterly* Volume 19, No. 4.
[21] NFPA *Quarterly* Volume 22, No. 3.
[22] NFPA *Quarterly* Volume 23, No. 1.
[23] NFPA *Volunteer Firemen* Volume 7, No. 5.
[24] NFPA *Quarterly* Volume 31, No. 1.
[25] NFPA *Volunteer Firemen* Volume 8.

Chapter VI

[1] McElroy, James K. Chapter 9, *Fire and the Air War.* NFPA Boston, Massachusetts, 1946.
[2] NFPA *Firemen Magazine* Volume 27, April, 1960.
[3] Collier's Encyclopedia, Volume 17, Crowell-Collier Educational Corporation, 666 5th Avenue, New York, New York. 1972.
[4] NFPA *Volunteer Firemen* Volume 9, No. 2. February, 1942.
[5] Miller, Francis T. *Comparative History of World II.* Home Education Guild, New York, New York. 1945.
[6] *Ibid.*
[7] *Ibid.*
[8] *Ibid.*
[9] *Ibid.*
[10] NFPA *Quarterly* Volume 35, No. 3.
[11] Bond, Horatio. *Fire and the Air War.* NFPA Boston, Massachusetts. 1946.

[12] *Ibid.*
[13] *Ibid.*
[14] *Ibid.*
[15] *Ibid.*
[16] NFPA *Quarterly* Volume 6, No. 3.
[17] NFPA *Volunteer Firemen* Volume 10, No. 1.
[18] NFPA Special Report E 117
[19] NFPA *Volunteer Firemen* August, 1944.
[20] NFPA *Volunteer Firemen* September, 1945.

Chapter VII

[1] NFPA *Quarterly* Volume 40, No. 1.
[2] NFPA *Quarterly* Volume 40, Nos. 1 and 3.
[3] NFPA *Quarterly* Volume 45, No. 2.
[4] NFPA *Quarterly* Volume 4, No. 3.
[5] NFPA *Quarterly* Volume 41, No. 1.
[6] NFPA *Quarterly* Volume 43, No. 1.
[7] *Ibid.*
[8] *Ibid.*
[9] NFPA *Firemen Magazine,* May, 1953.
[10] NFPA *Quarterly* Volume 49, No. 2; NFPA *Firemen Magazine,* October, 1955.
[11] NFPA *Quarterly* Volume 50, No. 2; NFPA *Firemen Magazine,* September, 1956.
[12] NFPA *Quarterly* Volume 53, No. 3; NFPA *Firemen Magazine,* January, 1959.
[13] NFPA *Firemen Magazine,* August, 1959.
[14] NFPA *Firemen Magazine,* March, 1956.

Aircraft and Airports

[1] Special Aircraft Accident Bulletin Series 1951; No. 2.
[2] Special Aircraft Accident Bulletin Series 1958; No. 8.
[3] Special Aircraft Accident Bulletin Series 1951; No. 2.
[4] Special Aircraft Accident Bulletin Series 1951; No. 1.
[5] Special Aircraft Accident Bulletin Series 1959; No. 1.
[6] Special Aircraft Accident Bulletin Series 1948; No. 37.
[7] NFPA *Quarterly,* October, 1962 pp. 101–110.
[8] NFPA *Fire Journal,* November, 1965, p. 17.
[9] NFPA *Fire Journal,* September, 1966 pp. 5–10.

Chapter VIII

[1] Willard, John N. *We Reach the Moon.* The New York Times Company. 1969.
[2] Bugbee, Percy. *Men Against Fire.* NFPA, Boston, Massachusetts. 1971.
[3] *Ibid.*
[4] NFPA Fire Command! January, 1972.
[5] NFPA Fire Command! September, October, 1967. NFPA Fire Journal, Volume 6, No. 2, 1966.
[6] NFPA Fire Journal, Volume 6, No. 2.
[7] NFPA Fire Journal, Volume 64, No. 2.
[8] NFPA Fire Command! April, 1972.
[9] NFPA No. SPP-25 — *Fires in High-Rise Buildings.* 1974.

Epilogue —

[1] NFPA Fire Protection Handbook, Fourteenth Edition.
[2] Based in part on an unpublished paper by Horatio Bond, Fire Protection Consultant, Hyannisport, Massachusetts.

Index